PROMOTING FLUOROCARBON LIFE CYCLE MANAGEMENT THROUGH CARBON MARKETS

DECEMBER 2024

Notes:
In this publication, "$" refers to United States dollars.
ADB recognizes "Siam" as Thailand.

On the cover: Hydrofluorocarbon recovery facilities (photo sourced from Creative Inc., Saitama, Japan).

Cover design by Josef Ilumin.

CONTENTS

TABLES, FIGURES, AND BOXES

FOREWORD

Increasing greenhouse gas (GHG) emissions have led to a global call for action in mitigating the effects of climate change, particularly in the Asia and Pacific region that is extremely vulnerable to disasters. From unprecedented droughts to devastating floods, the impacts of climate change in the region are palpably evident. Accelerating climate action to meet the Paris Agreement goals is thus an urgent priority that requires strategic and sustained global cooperation.

The past decade has seen tremendous efforts in mitigating emissions of carbon dioxide, which is a potent GHG. However, there is another GHG that poses a global warming impact that is thousands of times greater than carbon dioxide, namely hydrofluorocarbons (HFCs). Initially valued for its utility in refrigeration and air-conditioning, HFC emissions is now a major target for mitigation. The Kigali Amendment to the Montreal Protocol was precisely adopted to catalyze global efforts to curb HFC consumption and promote the reduction of gases emitted throughout the HFC life cycle. As countries around the world combat GHG emissions, the phasedown of HFCs has become a key component of their strategies for climate mitigation.

Underscoring its commitment to be Asia and the Pacific's climate bank, the Asian Development Bank (ADB) intends to deliver at least $100 billion in climate finance between 2019 and 2030. To better combat climate change and the other challenges constituting the region's polycrisis, ADB has updated its Capital Adequacy Framework and unlocked $100 billion in new funding capacity over the next decade. This expansion of available funds will be further leveraged by mobilizing private and domestic capital to scale the billions raised thus far to the trillions required to tackle the climate crisis. Through these initiatives, ADB seeks to foster sustainable development and environmental protection by delivering more programs and projects that address climate impacts, improve disaster readiness, and promote green infrastructure.

The development of high-integrity carbon markets is a critical component of ADB's holistic efforts in tackling climate change. If well-designed and effectively regulated, carbon markets function as critical pillars of the broader climate policy architecture that countries can adopt to mitigate climate change and enable their transition to low-carbon economies. For tackling HFC emissions, international carbon markets provide a cost-effective pathway as a market-based solution that can be leveraged to drive HFC mitigation.

This report makes a comprehensive case for the development of high-integrity carbon markets that will strategically advance the phasedown of HFCs. While the tasks required to implement these pioneering initiatives are filled with challenges, they are also brimming with opportunities for transformative change. This new interest in HFCs pertaining to Article 6 of the Paris Agreement is already opening doors that could lead to raising the mitigation ambitions of ADB members. With collective resolve, the urgent need for climate action vital to our sustainable future will be boosted by the inclusion of HFC mitigation in carbon markets.

As called for in the bank's Climate Change Action Plan, ADB is fully committed to supporting its members to enhance climate action through the effective implementation of strategies and initiatives best suited for national contexts. Working with our partners to ensure that economic aspirations align with environmental responsibilities, I believe that ADB and its members will contribute consequentially to fostering a healthier planet for generations to come.

Bruno Carrasco
Director General
Climate Change and Sustainable Development Department
Asian Development Bank

PREFACE

The use of hydrofluorocarbons (HFCs) has increased rapidly because of increasing global demand for air conditioners and refrigerators. Global baseline HFC emissions are expected to reach between 4.0 and 5.3 gigatons of carbon dioxide equivalent per year by 2050, 40%–58% of it from refrigeration and 21%–40% from standing air-conditioning; 51% of all HFC emissions are expected to be from Asia.

Recognizing the threat HFCs pose, countries adopted the Kigali Amendment to the Montreal Protocol in 2016, agreeing to gradually reduce the use of HFCs by more than 80% by 2047. Despite the international commitments in place, many countries have no laws or policies to curtail HFC emissions. Although the phasedown of HFCs is critical for achieving the temperature targets under the Paris Agreement, only a few countries have systems in place to enforce regulation and provide incentives for the proper management of HFCs.

Carbon markets play a critical role in mobilizing carbon finance to catalyze investments in low-carbon technologies and solutions and enhancing climate action. Carbon markets therefore have immense potential to play an integral role in the cost-effective phasedown of HFC emissions.

The Asian Development Bank (ADB) has been working at the forefront of the global carbon market development and leverages a long-standing engagement with carbon markets through its Carbon Market Program. ADB provides access to carbon finance to catalyze investments in low-carbon technologies and solutions and enhance climate mitigation actions among its developing member countries through a two-pronged approach—provide technical and capacity-building support and mobilize carbon finance.

Through the Carbon Market Program's Technical Support Facility, the technical assistance project on fluorocarbons life cycle management has provided Maldives, Mongolia, the Philippines, and Viet Nam with technical and capacity-building support to promote life cycle management of fluorocarbons, and energy efficiency in the cooling industry. Cooperation activities have included policy analysis, technology and business model survey, workshops for policymakers and mechanics, identification of appropriate technologies and investment projects, and provision of policy recommendations to each country. This publication is an output of the technical assistance, stemming from a common question across these countries on how life cycle refrigerant management can be financed and what is the role of carbon markets in HFC mitigation.

This publication delves into how carbon markets can facilitate the phasedown of HFC, and how carbon financing mechanisms can be leveraged to incentivize HFC mitigation actions in a manner that aligns with individual countries' broader economic, environmental, and sustainable development objectives. The publication discusses the potential rewards and challenges of participating in carbon markets at length and emphasizes the importance of strategically developing projects based on national needs and contexts by providing step-by-step guidance on how to identify and design them.

Overall, this publication will help policymakers increase their understanding on how carbon markets can be designed to enable cost-effective phasedown of HFCs, and the ways carbon markets can be structured to suit national climate objectives.

ACKNOWLEDGMENTS

This publication, *Promoting Life Cycle Management of Fluorocarbons Through Carbon Markets*, has been developed by the technical assistance project on fluorocarbons life cycle management of the Asian Development Bank (ADB) under its Carbon Market Program within its Climate Change and Sustainable Development Department (CCSD).

Virender Kumar Duggal, principal climate change specialist, CCSD, conceptualized and guided the development of this knowledge product. Ayse Frey and Jorge Luján Córdoba from Energy Changes conducted the research and produced the initial report, which is duly acknowledged and appreciated. The technical inputs from a team of ADB experts including Tatsuya Yanase, Akihiro Tamai, Jan Daniel Basco, Rastraraj Bhandari, Takahiro Murayama, Feby Utari, Anjum Choudhury, and Johan Nylander, as well as Mark D. Johnson from Ricardo Energy are also appreciated. This publication has hugely benefited from the peer review conducted by Priyantha Wijayatunga, senior director of ADB's Energy Sector Office.

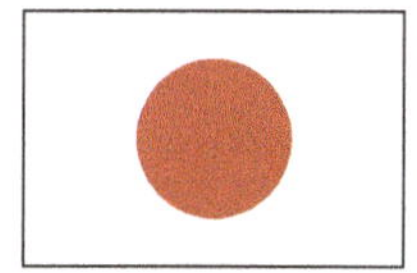

We would like to acknowledge the financial support from the People of Japan and the Asia Clean Energy Fund which fosters clean and efficient energy practices to reduce carbon emission in the region. Appreciation is also attributed to Manabu Takami, financing partnerships specialist and Rebecca Canoy of ADB's Partner Funds Division in supporting TA 6730-REG: Promoting Life Cycle Management of Fluorocarbons.

This publication was made possible by the valuable coordination and administrative support of Marya Laya Espaldon and Marilyn Parra. Cyrel San Gabriel edited this publication. Alvin Tubio did the layout and graphic design. Josef Ilumin designed the cover. Tuesday Soriano did the proofreading. Their diligent input is greatly acknowledged and appreciated.

ABBREVIATIONS

ACR	American Carbon Registry
ADB	Asian Development Bank
BTR	Biennial Transparency Report
CDM	Clean Development Mechanism
CFC	chlorofluorocarbon
CMA	Conference of the Parties serving as the Meeting of the Parties to the Paris Agreement
COPA	Climate and Ozone Protection Alliance
CORSIA	Carbon Offsetting and Reduction Scheme for International Aviation
EPR	extended producer responsibility
GHG	greenhouse gas
GIZ	Deutsche Gesellschaft für Internationale Zusammenarbeit
GWP	global warming potential
HCFC	hydrochlorofluorocarbon
HFC	hydrofluorocarbon
IPCC	Intergovernmental Panel on Climate Change
ITMO	internationally transferred mitigation outcome
LRM	life cycle refrigerant management
MLF	Multilateral Fund for the Implementation of the Montreal Protocol
MRV	monitoring, reporting, and verification
NDC	nationally determined contribution
ODS	ozone depleting substance
tCO$_2$e	tonnes of carbon dioxide equivalent
TEAP	Technology and Economic Assessment Panel
UNDP	United Nations Development Programme
UNFCCC	United Nations Framework Convention on Climate Change
US	United States
VCM	voluntary carbon market
VVB	validation and verification body

EXECUTIVE SUMMARY

Purpose, Scope, and Audience

This publication explores how countries may achieve progress toward emission reduction of fluorocarbons, most notably hydrofluorocarbons (HFCs), through strategically leveraging carbon markets, with focus on life cycle refrigerant management (LRM). Recognizing the significant environmental impact of these gases, including climate and ozone impacts, this publication presents LRM as an approach to reduce HFC emissions and help achieve the goals of the Montreal Protocol. The implementation of LRM measures faces several challenges including economic barriers, technical capacity, and infrastructure needs.

The report emphasizes how carbon finance and carbon market projects may be leveraged to address several of those barriers, including addressing finance needs for implementing LRM activities. It highlights that in order to meet the global climate and ozone goals of the Paris Agreement and the Montreal Protocol, countries must expand the coverage of their nationally determined contributions (NDCs) to include HFCs, and deepen mitigation efforts addressing fluorocarbons, supported in part by leveraging carbon markets strategically to achieve such policy objectives.

By providing technical inputs and practical advice to address both challenges and opportunities, the report aims to support policymakers and other stakeholders seeking to find solutions to implementing LRM and fluorocarbon emission mitigation measures, through leveraging carbon finance. This executive summary provides an overview of the policy and technical insights covered by this report, beginning with a visual summary of the report structure and main messages for policymakers.

Structure and Main Messages for Policymakers

This publication views policymakers as central players in the deployment of carbon markets toward the implementation of LRM. In addition to broad policy recommendations and steps, the report also provides specific insights and details of developing and deploying technical solutions for LRM through carbon markets, such as information on project development and design. This information may also be useful for prospective national stakeholders already engaged in the fluorocarbon industry nationally and those with potential interest in the development and operationalization of fluorocarbon carbon market projects.

For policymakers, four main actionable steps and insights are explained in Part 2, preceded by an introduction explaining the context and background of the topic in Part 1. These areas are highlighted below as part of this executive summary and also correspond to the main chapters and sections of this report. In the detailed report, more information and context are included that may assist both policymakers and other stakeholders in understanding better the concept and context of carbon market deployment in the fluorocarbon field.

The four main actionable areas described in Part 2 of this report are as follows:

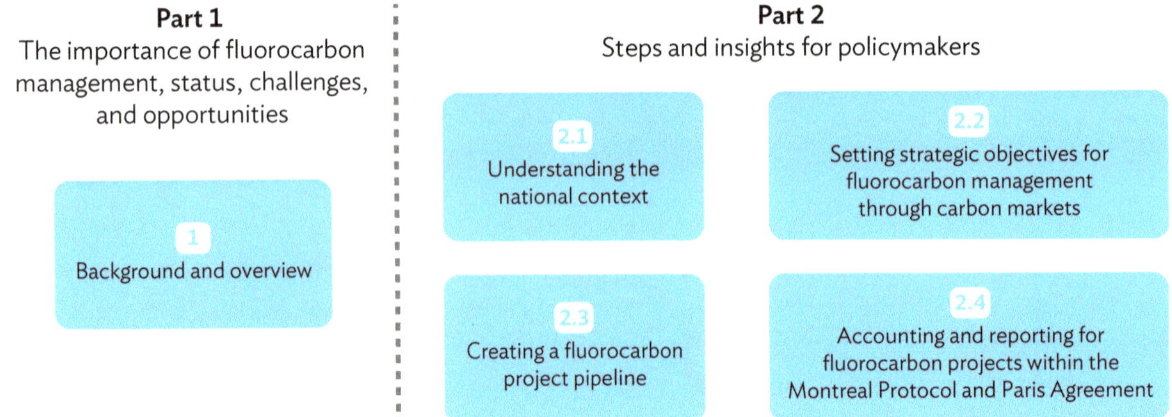

Climate Impacts of Fluorocarbons and Life Cycle Refrigerant Management

The Montreal Protocol has made remarkable progress in reducing the consumption of ozone depleting substances, and its Kigali Amendment has set targets for reducing HFC emissions. However, those fluorocarbons in use as refrigerants are largely uncontrolled in developing countries. LRM proposes an integrated strategy for managing these effectively and in a comprehensive manner by reducing emissions throughout the life cycle of fluorocarbons, such as by preventing leaks across the supply chain and enabling the effective recovery, reclamation, recycling, and destruction of fluorocarbons. LRM, if implemented effectively, can accelerate positive environmental outcomes and progress toward circular supply chains for fluorocarbons, and help countries meet their obligations under the Montreal Protocol and Paris Agreement. However, countries implementing LRM face several challenges, such as limitations in financing, infrastructure, and technical capacity, as well as complex supply chains.

Leveraging Carbon Markets for Fluorocarbon Management

Carbon markets, including both markets under Article 6 of the Paris Agreement and voluntary markets, hold the potential to address several of the challenges faced in LRM operationalization. To date, successful projects have addressed refrigerant emissions at various points in the supply chain, including supporting the destruction of fluorocarbons at end-of-life, leak prevention, or the reclamation of refrigerants. Deploying such projects strategically can support policy objectives as part of a comprehensive LRM strategy. However, implementing fluorocarbon projects in carbon markets also presents unique challenges and demands for countries hosting projects. Importantly, carbon credits must represent high-integrity emission reductions arising from projects that transparently implement mitigation activities, and countries must meet several requirements under the Paris Agreement. The report presents an overview of these requirements and practical advice for implementing effective carbon market fluorocarbon projects.

Understanding the National Context and Readiness for Carbon Market Participation

Understanding a country's national context is one of the first steps to strategically leverage carbon markets and support LRM efforts. Countries must assess their readiness toward participating in international carbon markets, including the requirements under Article 6 of the Paris Agreement. The following key national context and readiness items can guide countries in understanding their readiness status:

- **Evaluating national fluorocarbon management status.** Assessing the national status of fluorocarbon management nationally, including barriers and challenges to LRM approaches. This may include looking at data on fluorocarbon reclamation or destruction rates, relevant policies and regulations, and other fluorocarbon management approaches. Understanding these issues and the national context can support policymakers in choosing how to leverage carbon markets and the types of projects and policies that are most effective.

- **Determining readiness for international carbon market participation.** Understanding readiness steps for compliance with Article 6 requirements, including arrangements for authorizing internationally transferred mitigation outcome (ITMO) transfers, corresponding adjustments, and transparency provisions. As part of this task, policymakers may choose to fill readiness gaps with appropriate frameworks and legislation to meet requirements with Article 6 participation. This may take the form of Article 6 policies, regulations, or frameworks that are designed to meet international requirements and rules under Article 6.

- **Screening national climate policies and national inventory.** Reviewing the alignment of NDCs and the national inventory with requirements under Article 6 of the Paris Agreement. Based on the design of the NDC and national inventory, countries wishing to leverage carbon markets are strongly recommended to include HFC gases in the coverage of their NDCs and to report on HFCs when transacting ITMOs generated from HFC mitigation as part of their national inventory. These points are furthermore expanded in this executive summary and in the report.

- **Assessing infrastructure and capacity for fluorocarbon destruction and reclamation.** Policymakers can examine the availability of national facilities destroying or reclaiming fluorocarbons, as well as the readiness and capacity of public and private sector stakeholders, in order to establish further readiness gaps and needs.

Once these readiness items have been identified and examined, policymakers may take concrete steps to improve the status of readiness. This report provides an overview of the likely gaps and needs and examines how such items may be tackled. Based on an evaluation of national context, as described above, countries may also identify strategic objectives, such as increasing reclamation rates of fluorocarbons nationally, or incentivizing the proper disposal of end-of-life fluorocarbons, in addition to other LRM actions.

Creating a Pipeline of Fluorocarbon Projects

Stakeholders, both public and private, looking to leverage carbon markets for LRM implementation will also face the challenge of identifying high-quality projects and opportunities. The report provides an overview of current methodologies and key technical information around fluorocarbon project development relevant for policymakers and other stakeholders. When creating a pipeline of fluorocarbon projects, countries may leverage various approaches to identifying projects, such as carrying out a call for proposals, directly engaging with project developers or other national stakeholders, and engaging with buyers or multilateral institutions such as the Asian Development Bank. At this stage, different tasks will fall upon different stakeholders when designing and implementing projects. While policymakers may not be responsible for the tasks of identifying methodologies or designing specific technical aspects of projects, they may rely on project developers and other stakeholders to carry out these steps. The report provides information for interested policymakers and other stakeholders on these topics. Furthermore, different stakeholders, including policymakers, may be tasked with the options of finding buyers or establishing bilateral agreements for prospective buyers. The report provides information and actionable insights on identifying buyers from a multifaceted stakeholder perspective.

Accounting and Reporting of Fluorocarbon Projects

Robust accounting provisions are necessary for enabling transparency under the Paris Agreement and are essential in high-integrity projects. The report highlights the accounting and reporting provisions under Article 13 of the Paris Agreement, including the specific relevance of accounting and reporting for fluorocarbon projects, with specific recommendations. One primary actionable task is highlighted as an important task for policymakers and national governments: countries must report HFCs in their national inventory if transacting ITMOs arising from HFC mitigation and must include HFCs into their NDC in order to avoid overselling mitigation outcomes that are needed to achieve Paris Agreement targets.

Article 6.2 and Article 6.4 of the Paris Agreement

Article 6 of the Paris Agreement establishes how voluntary cooperation through international carbon markets can enable the achievement of nationally determined contributions (NDCs).

Article 6.2 provides a decentralized approach where countries can enter into bilateral or multilateral agreements for cooperation through the transfer of mitigation outcomes (emission reductions), known as internationally transferred mitigation outcomes (ITMOs).

The Article 6.4 market mechanism is seen as the successor of the Clean Development Mechanism. It is a centralized system under which tradable credit units are earned for reductions against baselines. It aims to promote the mitigation of greenhouse gases while fostering sustainable development. Credits may take the form of authorized emission reductions, which are ITMOs, or mitigation contribution units.

It is important to note that for ITMOs, the buying country is credited with the mitigation outcome. To implement this, a corresponding adjustment is required for each country's emission balances, covering any transfer (export) or receipt (import) of mitigation outcomes.

Irrespective of this corresponding adjustment, the host country takes advantage of the investment and co-benefits of the mitigation action.

Two Groups of Fluorocarbons

This publication categorizes fluorinated greenhouse gases (F-gases) or fluorocarbons into two main groups to be considered for greenhouse gas mitigation purposes and to align with international treaties. It uses the term "fluorocarbons" when referring to ozone depleting substances (ODS) and HFC substances. On the other hand, the term "refrigerants" is utilized in the context of refrigerant applications and includes other non-fluorocarbon refrigerant substances.

Ozone Depleting Substances

ODS are gases regulated by the Montreal Protocol on Substances that Deplete the Ozone Layer, which includes a phaseout plan for the production and consumption of specific ozone depleting gases such as chlorofluorocarbons, hydrochlorofluorocarbons, halons, and methyl bromide.

Hydrofluorocarbons

Hydrofluorocarbons (HFCs) are used as replacements for some of the ODS phased out by the Montreal Protocol. These gases are regulated by the Kigali Amendment, which was added to the Montreal Protocol in 2016. The amendment includes a phasedown schedule for the production and consumption of HFCs, which are also considered as a greenhouse gas under the Paris Agreement.

This distinction is relevant for leveraging the international carbon markets, as Article 6 of the Paris Agreement and many methodologies in the voluntary carbon market treat these gases separately.

PART 1

THE IMPORTANCE OF FLUOROCARBON MANAGEMENT, STATUS, CHALLENGES, AND OPPORTUNITIES

1.1 Climate Impacts of Fluorocarbons and Life Cycle Fluorocarbon Management

As global temperatures rise, the demand for cooling also grows, leading to increased refrigerant emissions that further contribute to global warming. Effective climate solutions and financing mechanisms are crucial for breaking this cycle.

Refrigerant emissions, from both ozone depleting substances (ODS), which include hydrochlorofluorocarbons (HCFCs) and chlorofluorocarbons (CFCs), and hydrofluorocarbons (HFCs) represent a climate threat with significant measurable negative impacts. Specifically, they cause (i) ozone depletion, measured in the ozone depleting potential of a substance; and (ii) global warming potential (GWP), representing the effect that fluorocarbons have on climate change. Fluorocarbons make up the majority of refrigerant substances that are utilized worldwide.[1] Depending on the specific type of fluorocarbon, they may have high ozone depleting potential and GWP, contributing to ozone depletion and climate change.

As a whole, the cooling sector has an outsized climate impact: some estimates put the emissions from the cooling sector at 13% of global greenhouse gas emissions by 2030, just over one-eighth of total emissions.[2] This makes up a significant part of global climate change and requires climate actions with technical solutions.

Countries have made progress in tackling fluorocarbon emissions through transparent, well-understood, and accurately measurable solutions and mitigation options. These technical solutions have already been piloted in a range of Asian Development Bank (ADB) developing member countries.

---

[1] United Nations Environment Programme. 2016. *Montreal Protocol on Substances that Deplete the Ozone Layer. Technology and Economic Assessment Panel Report.* Volume II. https://ozone.unep.org/system/files/documents/TEAP_ExIII-1 _Report_Sept-2016.pdf.

[2] Deutsche Gesellschaft für Internationale Zusammenarbeit (GIZ). 2019. *Naturally Cool! Green Cooling – for the Protection of the Climate and the Ozone Layer.* https://www.green-cooling-initiative.org/fileadmin/Publications/2019_Proklima _Panoramabogen.pdf.

The production and consumption of CFCs has been largely phased out by all countries, and developed countries are well advanced in the consumption reduction of HCFCs, with some having accomplished the phasedown. However, global fluorocarbon emissions continue to rise from existing equipment, end-of-life equipment, stockpiles, foams, and other products, particularly from HFCs.[3] By 2050, emissions from HFCs are projected to increase, and large banks of HFCs continue to increase in volume.[4]

While at more advanced stages of phasedown and phaseout, ODS emissions are also of great environmental importance, given the expected high share of retiring refrigeration and air-conditioning in the next decade being made up of ODS appliances and broad challenges in the management of fluorocarbons in general.

In many cases, countries are grappling with addressing fluorocarbon emissions at upstream, midstream, and downstream points, while implementing climate policies and meeting climate commitments. They must find solutions to enable the sustainable and effective management of fluorocarbons throughout their full life cycle, addressing a wide range of unique national circumstances and expectations. They are tasked also with achieving the Montreal Protocol obligations and implementing ambitious nationally determined contributions (NDCs). Implementing both requires adequate financing and resource allocation. Legal measures alone may not suffice to address the full ecosystem of fluorocarbon management. Furthermore, the implementation of solutions to address fluorocarbon emissions also requires capacity, technology transfer, and specialized expertise.

Countries are taking up the challenge: prior to 2020, only 28% of countries included HFCs in their NDCs. Post-2020, this number has increased to 47%.[5] However, this has not led to the widespread adoption of HFC management policies, except in a range of non–Article 5 countries, that is, developed countries needing to meet the advanced phasedown and control schedules of the Montreal Protocol. On the other hand, Article 5 countries (developing countries entitled to a 10-year delay in the compliance with control measures of the Montreal Protocol) are often at earlier implementation stages of such policies.

Globally, the implementation of policies and regulations comprehensively addressing fluorocarbons at all parts of their life cycle, including both ODS and HFCs, is concentrated in countries with already high awareness of fluorocarbon emissions. These countries also often have historical experience, understanding and consensus among stakeholders, and therefore are actively participating in implementation of policies.[6] While this is often not the case in Article 5 countries, progress is being made and awareness of the importance of fluorocarbon emissions and management is increasing.[7]

---

[3]   G. J. M. Velders et al. 2022. Projections of Hydrofluorocarbon (HFC) Emissions and the Resulting Global Warming Based on Recent Trends in Observed Abundances and Current Policies. *Atmospheric Chemistry and Physics.* 22(9). pp. 6087–6101. https://doi.org/10.5194/acp-22-6087-2022.

[4]   GIZ. 2012. *Global Banks of Ozone-Depleting Substances: A Country-Level Estimate 2.0.* https://www.giz.de/fach expertise/downloads/12%20ODS%20Global%20banks%20of%20ozone%20depleting%20substances%20-%20A%20 country-level%20estimate%202.0.pdf.

[5]   C. S. Malley et al. 2022. Integration of Short-lived Climate Pollutant and Air Pollutant Mitigation in Nationally Determined Contributions. *Climate Policy.* 23(10). pp. 1216–1228. https://doi.org/10.1080/14693062.2022.2125928.

[6]   A. Garg. 2023. *Activating Circular Economy for Sustainable Cooling: Global Best Practices on Lifecycle Refrigerant Management.* Council on Energy, Environment and Water. https://www.ceew.in/sites/default/files/global-best-practices-lifecycle-refrigerant-management-emissions.pdf.

[7]   United Nations Environment Programme. 2024. *Montreal Protocol on Substances that Deplete the Ozone Layer: Report of the Technology and Economic Assessment Panel. Decision XXXV/11 Task Force Report on Life Cycle Refrigerant Management.* https://ozone.unep.org/system/files/documents/TEAP-May2024-DecXXXV-11-TF-Report.pdf.

Given the impacts of these emissions, finding effective climate actions to tackle them can yield multiple benefits for countries. Indeed, the proper management and disposal of fluorocarbons support the accelerated achievement and implementation of the Montreal Protocol, the Kigali Amendment, and the Paris Agreement, among other co-benefits enabled by enacting effective policies.[8]

1.1.1 Life Cycle Refrigerant Management as a Novel Climate Solution

Historically, fluorocarbons have been an important environmental concern. Beginning with the discovery of the hole in the ozone layer, and the pinpointing of ODS as the cause, the issue has received considerable international attention. The Montreal Protocol regulated ODS and led to the adoption of HFCs as the "next generation" of refrigerant gases, the primary use of fluorocarbons. However, the high GWP of HFCs led to the Kigali Amendment to the Montreal Protocol, which aims to gradually reduce the production and consumption of HFCs. These regulations for fluorocarbons are driven in part by the need to finance historical transitions from older fluorocarbons to more modern alternatives. Today, the evolution of fluorocarbon management trends and new technologies have led to the development of interim solutions as well as modern green cooling systems that utilize other gases, such as natural refrigerants, with no ozone impacts and very low climate impacts.

In addressing the emissions arising from refrigerant gases, in particular fluorocarbons, life cycle refrigerant management (LRM) has gained considerable international attention. LRM refers to the comprehensive strategies set out to reduce emissions from fluorocarbon banks, including the upstream, midstream, and downstream stages. LRM may start with the prevention of leaks from design, manufacturing and assembly, servicing, and use of equipment utilizing fluorocarbons. At middle and downstream stages, it includes the recovery, reuse, recycling, reclamation, and destruction of fluorocarbons, as well as the infrastructure, networks, and technicians that will enable effective fluorocarbon management and the reduction of emissions.[9]

Different specific fluorocarbons, national contexts, and other circumstances impact how LRM may be adopted. Policies for LRM can include a range of measures such as integration with voluntary programs, compliance credit generation, incentive programs, and corporate programs.[10]

LRM holds significant importance and potential. If implemented effectively, it can enable circularity of existing fluorocarbons and lead to modern, state-of-the-art refrigerants being utilized nationally; energy-efficient appliances; skilled technicians being trained; comprehensive monitoring, reporting, and verification (MRV) systems being implemented; and smart fluorocarbon gas collection and destruction networks being put in place. Overall, the importance of LRM is gaining ground in global conversations around fluorocarbon management and climate action.

[8]   United Nations Environment Programme. 1987. *The Montreal Protocol on Substances that Deplete the Ozone Layer.* https://ozone.unep.org/treaties/montreal-protocol; United Nations Framework Convention on Climate Change. 2015. *The Paris Agreement.* https://unfccc.int/process-and-meetings/the-paris-agreement.

[9]   G. J. M. Velders et al. 2022. Projections of Hydrofluorocarbon (HFC) Emissions and the Resulting Global Warming Based on Recent Trends Observed Abundance and Current Policies. *Atmospheric Chemistry and Physics.* 22(9). pp. 6087–6100. https://doi.org/10.5194/acp-22-6087-2022.

[10]   C. S. Malley et al. 2022. Integration of Short-Lived Climate Pollutant and Air Pollutant Mitigation in Nationally Determined Contributions. *Climate Policy.* 23(10). pp. 1216–1228. https://doi.org/10.1080/14693062.2022.2125928.

1.1.2 Challenges in Mobilizing Life Cycle Refrigerant Management

Countries face several challenges in reducing fluorocarbons emissions in order to achieve their targets under the Montreal Protocol and the Paris Agreement. These include the crucial issue of raising finance that enables solutions to be implemented on the ground. To help raise finance for the implementation of life cycle fluorocarbon management activities, the Montreal Protocol provides financing through its Multilateral Fund for projects and initiatives for the phaseout of ODS and help countries meet compliance obligations. However, the scale of finance required to support the complete life cycle management of fluorocarbons is pressing, and further finance is needed to comply with obligations and implement LRM approaches effectively.

Challenges in Mobilizing Finance and Implementing Life Cycle Refrigerant Management

There are several challenges to effective fluorocarbon life cycle management. The Technology and Economic Assessment Panel (TEAP) identifies a range of challenges to LRM implementation, among them economic and financial ones (footnote 7). TEAP is an advisory body under the Montreal Protocol tasked with providing technical and economic advice to support the objectives of the Protocol, including identifying barriers to LRM implementation.

Except for a number of HFC destruction projects at production sites under the Clean Development Mechanism (CDM), mobilizing finance is a central challenge due to lack of legal and regulatory framework and domestic budget for waste management of these substances and low price of bulk refrigerants against reclamation. Adequate financing must be provided for infrastructure needs, such as cylinders for fluorocarbon collection, transport, recovery, reclamation or destruction, vehicles, labor, warehouses and locations, and other associated costs. These costs negatively impact the standalone economic viability of companies with a potential role to play in effective LRM, which results in the need for additional means of financing being required for LRM deployment.

These costs begin with upstream stages, such as leak prevention in the design, manufacturing, and assembly of equipment, and continue all the way through the use phase to recovery and reuse or destruction. They may also vary depending on a range of factors, including activity or project type, the specific fluorocarbon, technologies employed, and the accessibility of LRM elements. These costs are associated with steps such as

- leak detection and prevention;
- transport and storage;
- training of skilled technicians;
- establishing robust monitoring, reporting, and verification (MRV) systems;
- good equipment maintenance practices;
- recovery and collection of gases;
- testing laboratories qualified to perform chemical analyses to identify refrigerant composition for determining further processing steps, and for establishing whether processed refrigerants meet required product specifications;
- switching/modifying equipment to low-GWP refrigerants;
- transition to natural refrigerants;
- technologies for recovery, recycling, reclamation, and destruction;
- infrastructure development; and
- proper end-of-life disposal.

Furthermore, the costs and market situation of virgin fluorocarbons also have an impact on LRM implementation. Generally, high fluorocarbon prices have been associated with increased leak prevention, fluorocarbon recovery, and reuse in different contexts. However, high prices for reclaimed fluorocarbons are associated with the complex tasks of establishing collection networks, training technicians, and reclaiming costs, among others; while low prices of virgin refrigerants are a persistent barrier to LRM implementation in various national contexts. The latter case is particularly relevant for Article 5 countries at earlier stages of HFC phasedown.

Noneconomic Barriers to Life Cycle Refrigerant Management

Not all challenges that countries face when attempting to reduce their emissions from fluorocarbons are based on costs or mobilizing finance. For instance, transitioning to newer low-GWP technologies involves adopting more complex systems along with higher environmental, energy, and safety standards. This transition requires not only more technicians but also extensive retraining of the existing workforce. This may be influenced by the size of companies that are involved. While larger, international companies may have more means to build a skilled workforce in response to policy change or initiatives to promote LRM, smaller and midsize stakeholder companies may have less means to do so, and may lack the capacity or technical knowledge to make such transitions in general.

The TEAP identifies the sheer number of stakeholders to be engaged as a distinct barrier when designing and implementing LRM (footnote 7). Alongside the vast number of stakeholders involved in full-scale LRM are the complexity of supply chains. The networks surrounding the life cycle of fluorocarbons are often international, and as such touch upon international trade and corporations. Depending on national circumstances, countries pursuing LRM will have to engage with such networks and stakeholders in order to establish comprehensive policies. Other barriers include challenges with modern technology diffusion and uptake, market perceptions of reclaimed fluorocarbons, lack of technical implementation capacity, and absence of tracking and MRV systems for emissions that are in line with both the Montreal Protocol and the Paris Agreement.

In addition to this range of challenges, countries are also facing implementation costs for climate policies in general. As NDCs become more and more ambitious over successive NDC cycles, the costs of implementation are similarly expected to go up—some estimates place the cost of implementing conditional NDCs between $97 billion and $191 billion by 2030.[11] Accessing and deploying finance toward other measures included in an NDC, in other sectors, may compete with the finance needs of LRM, even when there is a strong rationale and willingness to implement LRM measures.

1.1.3 Opportunities and Benefits in Accelerating LRM Implementation

Once barriers are surpassed, significant opportunities may be leveraged. The Carbon Containment Lab lists several benefits that LRM offers, including important positive climate impacts.[12] Effective LRM will displace virgin production of fluorocarbons and help address other sources of emissions, such as leaks, venting, and other end-of-life emissions, resulting in a strong positive effect for the atmosphere, and in the case of ODS, for the ozone layer as well. Fluorocarbons that are not leaked and are reclaimed will support circular economy models and improve resource conservation (in fluorocarbons themselves) as well as the energy efficiency of appliances.

---

[11] A. F. Hof et al. 2017. Global and Regional Abatement Cost of Nationally Determined Contributions (NDCs) and of Enhanced Action to Levels Below 2°C and 1.5°C. *Environmental Science and Policy*. Volume 71. pp. 30–40. https://doi.org/10.1016/j.envsci.2017.02.008.

[12] C. Mayhew, T. Chao, and A. O'Rouke. 2023. Lifecycle Refrigerant Management: Maximizing the Atmospheric and Economic Benefits of the Montreal Protocol. Background Paper. Yale Carbon Containment Lab. https://carboncontainmentlab.org/documents/yale-cc-lab-lrm-background-paper-(updated-for-mop-35).pdf.

While strong architectures are in place in several countries establishing quotas and managing the existing stocks and imports of fluorocarbons, the capacity built up through LRM approaches will further result in a better national understanding and capacity development of emission sources and bank management. This may be in the form of more accurate inventories, MRV systems, and detailed data. Policymakers can then leverage this better data to make informed decisions about fluorocarbon management and use the lessons learned in operationalizing LRM in other sectors of the economy.

The implementation of LRM elements can lead to other positive benefits and opportunities. A major opportunity is the establishment of a reclaimed fluorocarbon market. In view of constraining HFC supplies due to the implementation of the Kigali Amendment, a reclaimed fluorocarbon market may help address price fluctuations in fluorocarbons nationally. The implementation of LRM also has other economic benefits, including job creation, training of workforce, and technology transfer.

1.2   Why Leverage International Carbon Markets to Manage Fluorocarbons: Opportunities and Challenges

To address some of the challenges for comprehensive LRM, carbon markets emerge as a potentially transformative tool that can, in a targeted manner, finance key points of an LRM strategy. As a mechanism to raise climate finance, carbon markets have been leveraged in the past in some jurisdictions to finance fluorocarbon emission reductions in a results-based, independently verified way. Carbon markets may also be leveraged strategically at key points of an LRM strategy to address the need for creating incentives, raising finance, or meeting other LRM strategic objectives.

This section introduces the international carbon markets, including Article 6 and voluntary carbon markets (VCMs), and discusses the rationale for leveraging carbon finance in order to operationalize LRM.

While carbon finance can be an effective tool to address fluorocarbon emissions and offers significant opportunities, there are also challenges involved in implementing these solutions. This analysis summarizes these challenges and explores ways to mitigate them.

1.2.1   What Are International Carbon Markets?

Carbon markets are market-based instruments that put a price on greenhouse gas emissions with the aim of reducing them cost-effectively. They function by issuing tradable credits based on the results of emission-reducing activities, which may then be sold to a variety of market participants, resulting in so-called carbon finance.

As part of the Montreal Protocol, successful targets and policies were set up to regulate the phaseout of the production and use of fluorocarbons. Carbon markets, to a limited extent, have been utilized as a financing instrument to address fluorocarbon emissions and to make progress toward those targets, especially in addressing end-of-life emissions, where fluorocarbons may be stockpiled or held in banks and slowly leak into the atmosphere.

International carbon standards have methodologies in place for the development and implementation of carbon projects addressing fluorocarbons, with a wide range of projects in multiple locations receiving carbon finance. For example, four carbon standards currently have methodologies and a range of registered projects destroying ODS: Verra, California Air Resources Board, Climate Action Reserve, and the American Carbon Registry. These methodologies feature detailed approaches on how projects must be designed and implemented—setting out rules, methods, and variables that project developers must follow in order to receive carbon credits.

Projects often feature various approaches to addressing the challenges of fluorocarbon emissions at upstream, midstream, and downstream points. These include, for example, operating facilities destroying or reclaiming fluorocarbons; training technicians in carrying out collection, leak repair, replacement, or other activities; and establishing networks or software tools for the collection and tracking of fluorocarbons, from source to end point. In addition, carbon market projects can also help bridge finance gaps at other points, such as financing a transition to more energy-efficient equipment or promoting natural or next-generation refrigerants, with pilot projects being developed for such applications in addition to end-of-life measures. Carbon markets, in this context, have demonstrated that they can play a role in providing financial incentives for the costly and otherwise unprofitable actions that make up a comprehensive LRM.

Carbon Markets at a Glance

There are two main types of carbon markets: compliance and voluntary. Compliance markets are leveraged to meet the requirements set out by regulations, laws, or international agreements, such as the Paris Agreement. They may include both emission trading schemes and crediting markets, which are the focus of this report. On the other hand, voluntary markets, which also often employ crediting approaches, are used for voluntary purposes, such as the fulfillment of voluntary climate commitments.[13]

Historically, these two markets have always been intertwined. The interplays and overlaps between them have been increasing over time, and it is expected that they will continue to converge.[14] An example of this is the use of the Certified Emission Reductions from the Clean Development Mechanism of the Kyoto Protocol for voluntary purposes. Another prominent example of the convergence of voluntary and compliance markets is the Carbon Offsetting and Reduction Scheme for International Aviation (CORSIA) mechanism adopted by the International Civil Aviation Organization to address international aviation emissions. CORSIA allows regulated airlines to use certain credits from the VCM, with a corresponding adjustment, for compliance with targets under the treaty.[15]

Utilizing these carbon market mechanisms, countries can enable strategic LRM through the implementation of smart policies and leveraging high-integrity carbon market project development. Generating revenues from the crediting of mitigation of fluorocarbons is a powerful tool for countries looking to enable these climate solutions in cost-effective ways. For example, carbon market mechanisms, including the mechanisms supported by Article 6 of the Paris Agreement and the VCM may serve to lower the implementation costs of the Paris Agreement by as much as 30% by 2030 and 50% by 2050, according to some estimates.[16] By extension, they also play a role in accelerating the objectives of the Montreal Protocol (Box 1).

[13] Asian Development Bank. 2023. *National Strategies for Carbon Markets Under the Paris Agreement: Making Informed Choices.* https://www.adb.org/sites/default/files/publication/928596/national-strategies-carbon-markets-paris-agreement.pdf.

[14] K. Wetterberg, J. Ellis, and L. Schneider. 2024. The Interplay Between Voluntary and Compliance Carbon Markets: Implications for Environmental Integrity. *OECD Environment Working Papers.* No. 244. Organisation for Economic Co-operation and Development (OECD) Publishing. https://doi.org/10.1787/500198e1-en.

[15] International Air Transport Association. 2024. *Fact Sheet: CORSIA.* https://www.iata.org/en/iata-repository/pressroom/fact-sheets/fact-sheet---corsia/.

[16] IETA. 2019. *The Economic Potential of Article 6 of the Paris Agreement and Implementation Challenges.* https://elibrary.acbfpact.org/acbf/collect/acbf/index/assoc/HASH015c/639e80ab/63c6ef7b/b949.dir/The%20Economic%20Potential%20of%20Article%206%20of%20the%20Paris%20Agreement%20and%20Implementation%20Challenges.pdf.

> **Box 1**
>
> ## Leveraging Carbon Markets for Ozone Depleting Substance Destruction in Thailand
>
> **Background.** The project involves the destruction of ozone depleting substance (ODS) refrigerants, specifically CFC-12, which have been stockpiled by the Government of Thailand over decades. These highly potent greenhouse gases, of which production had been banned and were brought into the country without the necessary permits from the Government of Thailand, were collected and stockpiled by the Customs Department as part of its efforts to enforce the Montreal Protocol, resulting in more than 10,000 cylinders stored in dozens of depots and warehouses throughout the country. However, in the absence of laws or regulations requiring the destruction of ODS, and with limited financial resources to support these costly activities—especially given the lack of alternative revenue streams—these gases would remain stored in cylinders, gradually leaking over time.
>
> **Tradewater Carbon Project.** Tradewater, a United States-based company that specializes in the recovery and destruction of refrigerant gases, has collaborated with the Thailand Customs Department to find solutions to address this problem through the generation of carbon credits.[a] In this project, the voluntary carbon market is leveraged via the American Carbon Registry (ACR) methodology, "Destruction of Ozone Depleting Substances from International Sources," to generate carbon credits and thereby provide economic incentives for the recovery and permanent destruction of these gases in a local destruction facility, Waste Management Siam.[b] From 2022 to 2023, the first six phases of this project generated more than 1.1 million tonnes of carbon dioxide equivalent in carbon credits from avoided emissions, according to the information from the ACR.[c] Until destruction is mandated by law and other economic incentives are available to enable such end-of-life solutions, carbon markets have provided a financing mechanism for destroying greenhouse gas emissions that otherwise would have leaked into the atmosphere.
>
> **Sustainable Development Co-Benefits.** In addition to its climate impact, the project also contributes to sustainable development of the waste management industry in Thailand. It has created a proven process for destroying CFC-12 gases. The project has also created jobs and provided training, established the supply chain logistics, and ensured the Waste Management Siam incinerator meets the process requirements of the Montreal Protocol and reaches the necessary high temperatures required to destroy 99.99% of the CFC-12 gases. The project has met the ACR's requirements for sampling of the gases with third-party verification, conducted continuous monitoring with regular intervals of reporting, and followed all regulatory and environmental rules.
>
> [a] ACR. 2024. *Complementing the Montreal Protocol: The Tradewater – Thailand Carbon Project.* https://acrcarbon.org/wp-content/uploads/2024/09/ACR-Case-Study-Tradewater-in-Thailand-FINAL.pdf.
>
> [b] ACR. 2021. *Destruction of Ozone Depleting Substances from International Sources.* https://acrcarbon.org/methodology/destruction-of-ozone-depleting-substances-from-international-sources/.
>
> [c] ACR. 2024. *A Trusted and Transparent GHG Registry.* https://acrcarbon.org/acr-registry/.

Of key importance to leveraging carbon markets effectively, both under Article 6 and in the VCM, is the need to ensure environmental integrity through carefully designed project methodologies and a range of risk mitigation measures. These practices ensure that emission reductions are real, measurable, and permanent, resulting in higher credibility and carbon prices and enabling greater climate impact. Creating conditions for high-integrity projects should be a policy priority resulting in effective and transparent climate solutions.

On the other hand, low environmental integrity poses several negative consequences, underpinning the need for effective design and implementation of carbon projects. A potential consequence of low-integrity carbon market projects is the potential for failure to effectively achieve climate targets, exacerbating climate change instead of tackling it. In the case of ODS, low-integrity carbon market projects can have negative impacts on the ozone layer. In addition, even a small number of low-integrity carbon projects may have a huge impact on the public perception and credibility of carbon finance, including LRM, as a mechanism to drive sustainable climate actions as a whole. Therefore, maintaining high integrity as a priority in leveraging carbon finance for fluorocarbon management and LRM remains uniquely important.

1.2.2 How Can Carbon Credits Support Life Cycle Refrigerant Management?

LRM may be supported by leveraging carbon markets to provide finance and incentives through the crediting of emission reductions as part of carbon market projects, for key policies, measures, infrastructure, and stakeholders. Where gaps have existed in policy or financing solutions, carbon markets have been employed effectively in the past to support refrigerant management. For instance, in the United States (US), carbon credits have played a role in accelerating the adoption of advanced refrigeration systems, enabling food service providers to switch to low-GWP refrigerants faster.[17]

Given their experience in managing fluorocarbons as part of the Montreal Protocol, countries are often aware of the degree of complexity in tackling LRM. Intervention points for effective LRM may be categorized into upstream (production of refrigerants), midstream (in-use refrigerants), and downstream (end-of-life). Figure 1 provides an overview of the potential solutions that carbon markets offer to enable LRM. A comprehensive strategy may employ carbon market mechanisms to enable solutions for all or some of the upstream, midstream, and downstream challenges.

It is possible to carry out such solutions by developing various types of crediting projects. These solutions can therefore be a part of an overarching LRM strategy. To date, multiple methodologies have been developed in the VCM to design, quantify, audit, and certify the project types described. Through Article 6, countries may also define cooperative approaches with partner countries and define crediting projects where internationally transferred mitigation outcomes (ITMOs) are exchanged for mitigation arising from the solutions presented in Figure 1.

Such a strategy will vary depending on national circumstances and a wide range of factors, including national legislation or local policies in place to govern fluorocarbon management. Countries with no fluorocarbon destruction or reclamation facilities will face a different set of questions when implementing an LRM approach supported by carbon markets than those with access to such facilities. In their effort to leverage carbon markets, for instance, countries with no existing facilities for reclaiming or destroying fluorocarbons may choose to employ carbon finance to catalyze the construction of such facilities. Countries with already established facilities, on the other hand, may choose to focus carbon finance opportunities to establish or improve collection and transport systems and technician capacity, or modernize fluorocarbon applications and promote technology transfer.

Countries may also want to prioritize a certain category of fluorocarbon management depending on other national priorities, such as focusing on midstream solutions to enable higher energy efficiency and accelerating Kigali Amendment implementation and NDC achievement.

---

[17] M. Tiwari. 2023. *Cleaning Up Global Cooling Systems with Carbon Markets.* Winrock International. https://winrock.org/cleaning-up-global-cooling-systems-with-carbon-markets/.

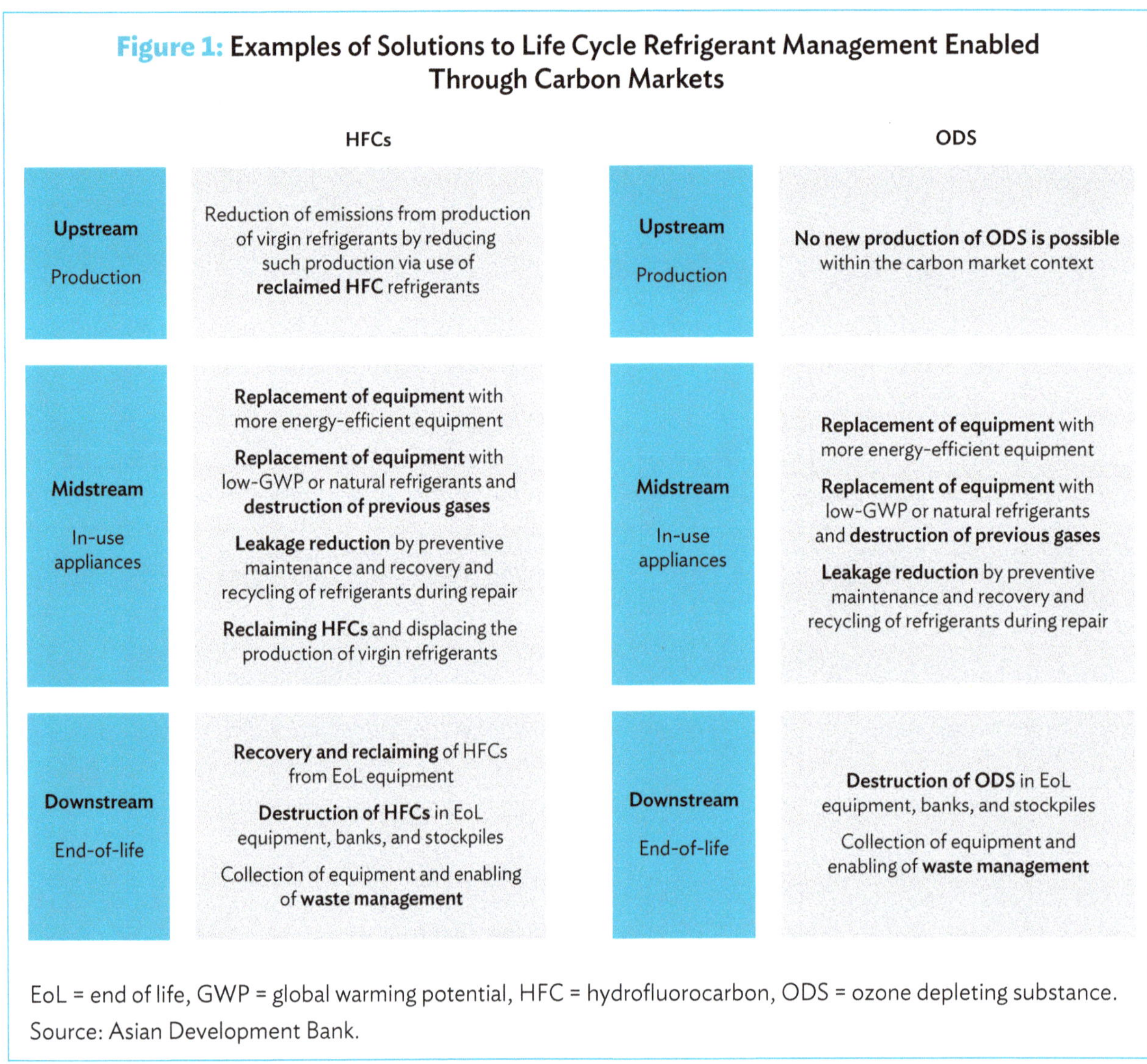

EoL = end of life, GWP = global warming potential, HFC = hydrofluorocarbon, ODS = ozone depleting substance.
Source: Asian Development Bank.

Overall, the type of activities that may be pursued nationally will vary based on national circumstances and context, as well as national strategy and policy that is being pursued to tackle fluorocarbon emissions in general.

1.2.3 Carbon Markets Under Article 6 of the Paris Agreement

Article 6 of the Paris Agreement enables countries to collaborate to achieve their NDC targets by leveraging carbon markets, creating market-based pathways for finance flows. It provides countries with rules, guidance, and procedures to establish carbon markets under the United Nations Framework Convention on Climate Change (UNFCCC) umbrella and to enable the exchange of credits under Article 6.2 and Article 6.4 (also known as the Paris Agreement Crediting Mechanism).

It relies on a variety of activities, including "traditional" carbon market projects that can generate carbon credits. If authorized by the host country and if they meet several criteria under Article 6, then the credits become ITMOs.

ITMOs may be used by countries to count toward achieving their NDC targets. In addition, ITMOs may be used for Other International Mitigation Purposes, such as CORSIA, or they may be purchased by other market actors for voluntary commitments.[18]

In addition to the development of project-by-project carbon market activities under Article 6, countries may also consider sector crediting or policy crediting approaches. Such examples of collaboration will credit, for example, the overachievement of mitigation targets within a particular sector, or credit the implementation of a policy and its mitigation impact.

Broadly, Article 6 carbon markets, both Article 6.2 and Article 6.4, are results-based climate finance instruments certifying and crediting a range of climate actions.

Corresponding Adjustments

Central to Article 6 is the concept of corresponding adjustments. Corresponding adjustments are an accounting procedure that transacting countries must carry out to prevent the double counting of ITMOs toward a variety of purposes. In a nutshell, a host country transferring ITMOs to a buyer country, must carry out a corresponding adjustment whereby it declares that it will not count the transferred ITMOs toward its NDC target. Meanwhile, a country buyer of the credits will perform the inverse and count the purchased ITMOs toward its NDC target.[19] Figure 2 shows how corresponding adjustments function for a host country and buyer country.

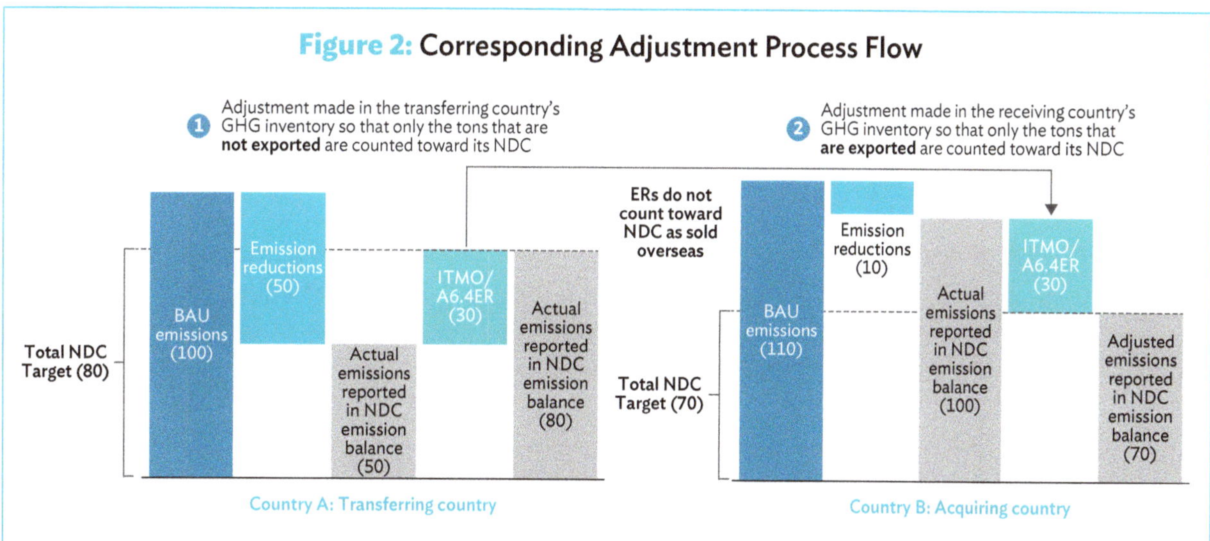

BAU = business as usual, ER = emission reduction, GHG = greenhouse gas, ITMO = internationally transferred mitigation outcome, NDC = nationally determined contribution.

Source: Asian Development Bank. 2023. *National Strategies for Carbon Markets Under the Paris Agreement: Making Informed Policy Choices.* https://www.adb.org/sites/default/files/publication/928596/national-strategies-carbon-markets-paris-agreement.pdf.

---

[18] United Nations Framework Convention on Climate Change (UNFCCC), Conference of the Parties serving as the meeting of the Parties to the Paris Agreement (CMA). 2021. Decision 2/CMA.3, Guidance on Cooperative Approaches Referred to in Article 6, Paragraph 2, of the Paris Agreement. https://unfccc.int/sites/default/files/resource/cma2021_10_add1_adv.pdf.

[19] UNFCCC, CMA. 2021. Decision 2/CMA.3. Annex, Paras. 7–10. Guidance on Cooperative Approaches Referred to in Article 6, Paragraph 2, of the Paris Agreement. https://unfccc.int/sites/default/files/resource/cma2021_10_add1_adv.pdf.

Corresponding adjustments may be carried out in a variety of metrics, based on the metrics included in countries' NDCs, such as tonnes of carbon dioxide equivalent (tCO_2e). While corresponding adjustments are possible in non-greenhouse gas (GHG) metrics, no current examples of such transactions have taken place to date. Non-GHG metrics may include ODS gases, which are not regulated under the Paris Agreement.

Progress to Date on Article 6

Article 6 is fully operational after the conclusion of COP29 in Baku, and the rules for Article 6.2 and for the Paris Agreement Crediting Mechanism under Article 6.4, are largely in place.[20, 21]

From the implementation side, countries are continuing to implement regulatory and institutional frameworks and are making steady progress toward signing bilateral agreements. They are developing the relationships needed between countries to collaborate in order to generate ITMOs and achieve their NDCs. To date, six countries have submitted initial reports, and 86 bilateral agreements have been made or are at various stages of maturity.[22]

Multiple projects are being implemented across Africa, Latin America, Southeast Asia, and West Asia; and more bilateral agreements are being signed.

Examples of current projects being credited as "typical" carbon market projects include cookstoves, agricultural practices improvement, waste recycling composting, solar renewable energy, and e-mobility. Policy crediting approaches are also being developed under Article 6. An example is policy crediting in Uzbekistan, where implementation and enforcement of policies that accelerate the clean energy transition by supporting energy efficiency measures, phasing out energy subsidies, and transitioning to cleaner energy sources are being planned to generate ITMOs under Article 6.2 of the Paris Agreement.[23] Both project-by-project crediting as well as policy crediting approaches are possible for addressing fluorocarbons through Article 6.

Linkages Between Montreal Protocol, Kigali Amendment, and Paris Agreement

A central theme in this document are the implementation challenges related to the different scope and coverage of the two landmark international treaties addressing fluorocarbons: the Montreal Protocol and the Paris Agreement. Addressing the implementation challenges begins with understanding the scope of these treaties.

Table 1 presents an overview of the differing coverage between treaties, with the Montreal Protocol covering the phaseout of ODS production and consumption, the Kigali Amendment providing for the phasedown of HFC production and consumption, and the Paris Agreement explicitly regulating only HFC emissions but allowing flexibility to report ODS mitigation targets.

[20]   UNFCCC, CMA. 2021. Decision 2/CMA.3, Guidance on Cooperative Approaches Referred to in Article 6, Paragraph 2, of the Paris Agreement. https://unfccc.int/sites/default/files/resource/cma2021_10_add1_adv.pdf.

[21]   UNFCCC, CMA. 2021. Decision 3/CMA.3, Rules, Modalities and Procedures for the Mechanism Established by Article 6, Paragraph 4, of the Paris Agreement. https://unfccc.int/sites/default/files/resource/cma2021_10_add1_adv.pdf.

[22]   United Nations Environment Programme. 2024. *Article 6 Pipeline.* https://unepccc.org/article-6-pipeline/.

[23]   World Bank. 2024. *Uzbekistan Receives $7.5 Million in Carbon Credits for Enabling Half a Million Tons of Emissions Reduction.* Press Release. https://www.worldbank.org/en/news/press-release/2024/06/21/uzbekistan-receives-7-5-million-in-carbon-credits-for-enabling-half-a-million-tons-of-emissions-reduction.

Table 1: Coverage of Montreal Protocol, Kigali Amendment, and Paris Agreement

	Montreal Protocol	Kigali Amendment[b]	Paris Agreement
Upstream (production)	ODS	HFCs	HFCs[a] ODS (optional)
Midstream (consumption)	ODS	HFCs	HFCs[a] ODS (optional)
Downstream (end-of-life)			HFCs[a] ODS (optional)

HFC = hydrofluorocarbon, ODS = ozone depleting substance.

[a] Only emissions are reported as part of a national inventory, not banks.

[b] There is no accounting or obligation for end-of-life disposal through the Kigali Amendment.

Note: The Montreal Protocol and Kigali Amendment regulate production and consumption of ODS and HFC gases whereas the Paris Agreement regulates emissions of greenhouse gases.

Source: Asian Development Bank.

While only a few projects involving end-of-life of fluorocarbons—that is, the destruction of the substances—have taken place under the Montreal Protocol, these have largely taken place in non-Article 5 countries and have destroyed fluorocarbons outside of national allowances, with only a small number of demonstration projects taking place in Article 5 countries.[24] Therefore, countries may still need solutions, support, and finance to address fluorocarbons and implement specific measures for LRM. In Table 1, "coverage" denotes accounting and reporting requirements and does not directly indicate financing solutions.

The different coverage of substances under each treaty has direct impact for countries addressing them. Signatories to both the Montreal Protocol and Paris Agreement have important accounting and reporting obligations, which are relevant when crediting projects through Article 6. To fulfill obligations under the Montreal Protocol, a National Ozone Unit must have a comprehensive oversight of the production and use of ODS and HFC gases in the country (with production being relevant only for very few countries). When accounting emissions under the Paris Agreement, countries develop National Inventory Documents (NIDs) through Intergovernmental Panel on Climate Change (IPCC) guidance, covering the gases included in the agreement.

As such, countries addressing fluorocarbon emissions must be able to establish accounting and reporting systems to be able to accurately measure their impact, particularly when leveraging carbon finance.

Challenges in the Eligibility of ODS Projects Under Article 6

The coverage of different substances under the Paris Agreement and the Montreal Protocol raises the question of how countries may employ Article 6 mechanisms for crediting ODS destruction activities, when they are not included under the Paris Agreement. Transacting ITMOs credited for ODS mitigation under Article 6 of the Paris Agreement, despite the significant opportunity that can help toward establishing LRM nationally, still faces some barriers unlike when transacting HFC ITMOs.

---

[24] Climate and Ozone Protection Alliance (COPA). 2023. *ODS/HFC Reclamation and Destruction Technologies: A Review for Article 5 Countries.* https://www.copalliance.org/imglib/publications/COPA_Publication_ODS_HFC_%20Destruction_Reclamation.pdf.

ODS, as mentioned previously, are not included under the Paris Agreement and therefore, are typically outside the scope of many NDCs, with only a few exceptions. As a central tenet of Article 6, corresponding adjustments are required for any activity transacting ITMOs, including whether an activity is outside of the scope of the NDC. This way, countries will be effectively accounting for the international transfer of emission reductions, without observing an associated emission reduction as part of their national inventory. While essential for the avoidance of double counting, this mechanism essentially creates an overselling risk and potentially impacts the achievement of the NDC target.[25] This challenge is known as the visibility of emission reductions.

This creates a disincentive for countries looking for ITMO-generating activities to be carried out in sectors or gases outside the scope of their NDC. Countries must determine whether the potential exchange of ITMOs under Article 6, with a matching corresponding adjustment, for projects mitigating ODS emissions, are sufficiently attractive and offer a sufficient incentive in the form of a carbon finance obtained in exchange for the ITMOs, and additional co-benefits to effectively make up a sufficient incentive for corresponding adjustment costs.

1.2.4 Voluntary Carbon Markets

Voluntary carbon markets (VCMs) have emerged in parallel with compliance markets and operate outside of mandatory compliance schemes, where participants are not formally obligated to achieve a certain target. Entities such as companies, organizations, or cities voluntarily purchase carbon credits to achieve voluntary mitigation targets. Historically, a VCM does not only complement compliance markets but also acts as a testing ground for financing new, innovative climate solutions that would otherwise not get the necessary attention or resources. They exist within an ecosystem of independent standard-setting organizations tasked with curating methodologies, implementing transparent processes for project development, and providing the overall framework for project development.

Refrigerant Projects in the Voluntary Carbon Market

Historically, most of the fluorocarbon projects developed under the VCM have been limited to credits generated in the US, where carbon finance has played a key role in catalyzing fluorocarbon life cycle management. Initially, voluntary buyers have shown an interest in purchasing credits. Meanwhile, a convergence has also taken place between voluntary markets and compliance markets, where regulatory programs such as the California Cap and Trade system allow the use of ODS destruction credits for compliance, thereby creating a demand for these credits, potentially high enough to incentivize ODS destruction.[26] As a result, private fluorocarbon management companies have been established, which have grown and ventured also into other project types such as HFC reclamation or switching to advanced refrigeration systems. In parallel, these companies were able to spend resources to improve demand for voluntary credits. By educating buyers and building awareness for these new, less-known project types, they were able to establish a voluntary market for fluorocarbon projects.

Beyond the US, the uptake of these project types has been slow for various underlying reasons. To date, 33 fluorocarbon gas projects in the VCMs have taken place outside of the US (Table 2). These projects have been either listed, registered, or are undergoing development or validation.

---

25   L. Schneider et al. 2022. Visibility of Carbon Market Approaches in Greenhouse Gas Inventories. *Carbon Management*. 13(1). pp. 279–293. https://doi.org/10.1080/17583004.2022.2075283.

26   California Air Resources Board. 2024. *Cap-and-Trade Program*. https://ww2.arb.ca.gov/our-work/programs/cap-and-trade-program.

Table 2: **List of Voluntary Projects in Fluorocarbon Management and Issued Credits**

Country	Number of Voluntary Carbon Market Projects (listed, registered, or undergoing development or validation)	Total Credits Issued as of October 2024 (tCO_2e)
Canada	13	908,257
Dominican Republic	1	23,657
France	3	61,861
Ghana	1	155,431
Mexico	6	1,686,395
Saudi Arabia	1	0
South Africa	2	0
Thailand	6	1,116,015
United States	429	76,952,116
Total		**80,903,732**

tCO_2e = tonnes of carbon dioxide equivalent.

Source: An elaboration of data from Berkeley Carbon Trading Project. 2024. Voluntary Registry Offsets Database.

1.2.5 Challenges in Addressing Fluorocarbon Emissions Through Carbon Markets

While carbon markets can be a powerful tool in catalyzing global decarbonization, there are also some challenges that need to be overcome for them to be successful and truly transformative. These can be categorized into supply-side challenges and demand-side challenges.

Supply-Side Challenges

Supply-side challenges are the challenges that countries will face when addressing fluorocarbon emissions through carbon markets and when obtaining carbon finance. These may cover issues such as identifying projects and applicable methodologies, cooperation between government sections, ensuring high-integrity credits, and meeting participation requirements. These are described in more detail below.

- **Identifying projects and applicable methodologies.** Identifying a project that could potentially be eligible for carbon credits is not a straightforward task, as is evident from a large number of consultancy firms active in this space. The complexity of various carbon market mechanisms makes it difficult to navigate and understand what types of fluorocarbon projects—under which conditions and under which crediting scheme and market—are eligible to generate credits.

  For high-integrity crediting, it is crucial to apply robust accounting methodologies that have ideally been approved by independent standards, UNFCCC, or another reputable framework. The number of fluorocarbon methodologies approved by independent standards or UNFCCC is currently very low and have limitations such as scope or geographic region. Furthermore, the requirements in these methodologies can be quite specific and/or stringent, making them difficult to apply to real-life cases. For example, the Technology and Economic Assessment Panel (TEAP) identifies requirements for destruction facilities, which often require significant modifications to existing equipment.

- **Cooperation between government sections.** Unlike many other project types in the carbon market, fluorocarbons represent a unique case where two international agreements regulate them (Montreal Protocol and Paris Agreement), necessitating a level of cooperation between government sections not typically required for most other project types. Specifically, this may involve a need of data for the elaboration of high-integrity baselines and the managing of accounting and an HFC inventory, while also tapping into knowledge about carbon market mechanisms and climate action policies, including Paris Agreement commitments and measures. These functions may be split into different government sections. For the former, experts in a National Ozone Unit will be knowledgeable in the national context and data on fluorocarbon management nationally, while in the latter, a different government unit is typically responsible for climate action involving carbon markets. Cooperation between these two units is recommended to establish institutional arrangements for managing administrative and data-driven processes where expert knowledge in both government units is tapped.

- **Ensuring high-integrity credits.** For a carbon market to be truly effective and deliver genuine climate impact, carbon credits need to ensure that emission reductions are real, additional, and verifiable and come with high environmental and social integrity. However, developing methodologies and projects with high quality and high integrity also comes at a higher cost and requires a deep understanding of challenges associated with setting robust baselines and estimating emission reductions conservatively. While a number of methodologies have been developed to date tackling fluorocarbon emissions, there is a clear need for further innovative and integrity-minded methodologies to be developed in the field.

- **Meeting Article 6 participation requirements, accounting provisions, and robust monitoring, reporting, and verification (MRV).** Countries must have in place a range of Article 6 requirements as described previously. In addition, they must establish robust MRV, create an inventory of relevant fluorocarbons, and build up the national capacity needed to implement such transparency systems. Overall, meeting these challenges will result in a strong benefit: climate transparency, national capacity, a good understanding of the management of fluorocarbons from an accounting and reporting perspective, and a deeper understanding of national challenges and status.

Demand-Side Challenges

The challenges on the demand side may be more straightforward to define, though not necessarily easier to address. Essentially, there must be sufficient demand for credits and a stable, reliable, and adequately high carbon price to ensure the viability of projects and give investors the confidence in this source of revenue. Historically, in both the compliance and voluntary markets, the demand and price for carbon credits has been unpredictable and inconsistent, making it necessary to take risks and obtain long-term offtake agreements where possible.

Broadly, Article 6 helps address demand-side challenges by establishing mechanisms that can be widely employed, with broad consensus of the signatories to the Paris Agreement and with links to compliance markets. Two key demand sources are linked to Article 6: NDC compliance and Carbon Offsetting and Reduction Scheme for International Aviation (CORSIA).

The VCM reached a peak value of $2 billion in 2021, and at that time, was projected to grow to more than $50 billion by 2030.[27] However, starting in 2022, increased media scrutiny and newspaper investigations challenged the quality and accuracy of the calculations underlying certain types of carbon credits.

---

[27] C. Blaufelder et al. 2021. A Blueprint for Scaling Voluntary Carbon Markets to Meet the Climate Challenge. McKinsey & Company. 29 January. https://www.mckinsey.com/capabilities/sustainability/our-insights/a-blueprint-for-scaling-voluntary-carbon-markets-to-meet-the-climate-challenge.

Combined with the lack of regulation, this ultimately undermined trust in the VCM, resulting in a sharp drop in demand and prices by 2023. As a result, several initiatives have been established to address integrity issues and enhance credibility and trust in the market, such as the Integrity Council for the Voluntary Carbon Market, which aims to set standards and guidelines for high-quality carbon credits in the VCM.

The demand for fluorocarbon credits in the VCM has remained relatively low compared to other project types (except in the US), mainly because of a combination of a perceived lack of sustainable development co-benefits and reputational issues stemming from historic HFC-23 credits under the Clean Development Mechanism (CDM). Although demand is difficult to quantify, according to data from the Berkley Carbon Trading Project's Voluntary Registry Offsets Database, more than 951,250 carbon credits arising from fluorocarbon projects were retired or canceled between 2020 and 2024, outside of the US.[28]

Depending on market dynamics and the evolution of the public perception of carbon projects, VCM credit demand may shift in the short, medium, or long term. As compliance markets continue to be implemented and operationalized, alongside ongoing integrity initiatives in the VCM, trust in carbon markets as reliable financing tools for decarbonization pathways is expected to grow.

---

[28] Berkeley Carbon Trading Project. 2024. *Voluntary Registry Offsets Database.* https://gspp.berkeley.edu/research-and-impact/centers/cepp/projects/berkeley-carbon-trading-project/offsets-database.

PRACTICAL GUIDANCE FOR OPERATIONALIZING CARBON MARKETS FOR LIFE CYCLE REFRIGERANT MANAGEMENT

Part 2 of this report describes how countries can operationalize carbon markets toward life cycle refrigerant management (LRM) and the key steps to achieve this. These steps are structured around four key elements supporting decision-making related to national circumstances and readiness, which are to be considered when tackling the strategic management of fluorocarbons through carbon markets (Figure 3).

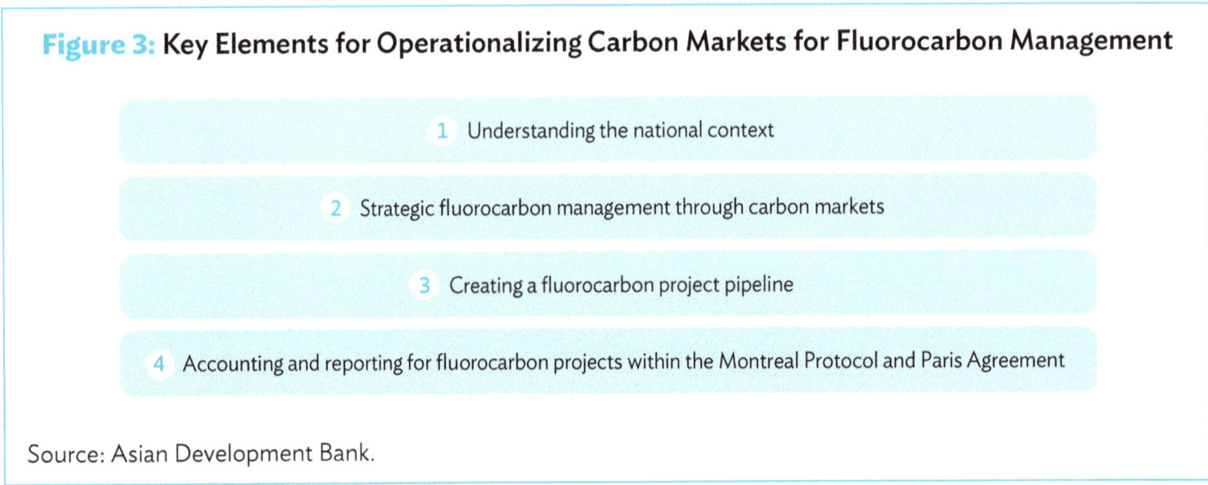

Figure 3: Key Elements for Operationalizing Carbon Markets for Fluorocarbon Management

Source: Asian Development Bank.

2.1 Understanding the National Context

This section presents how countries may better understand the national context of fluorocarbon management and evaluate their readiness for participating in carbon markets, including Article 6 of the Paris Agreement and voluntary carbon markets (VCMs). It is divided into several important questions that countries must be aware of when implementing carbon markets to address fluorocarbon life cycle emissions:

- What is the status of fluorocarbon management nationally?
- What are the sources of emissions of fluorocarbons nationally?
- What is the status of national readiness to participate in international carbon markets?
- What are the linkages of the national inventory and the NDC to carbon markets?

2.1.1 Evaluating the Status of Fluorocarbon Management Nationally

Amid a range of climate threats and competing priorities, having a clear picture of gaps and needs toward the deployment of carbon markets to address climate issues, such as addressing fluorocarbon emissions, is essential for countries beginning to consider these solutions. Understanding the national context, including the day-to-day issues of fluorocarbon management, can inform how carbon markets can be implemented effectively to achieve specific objectives.

Countries in Asia are at varying stages of fluorocarbon management (footnote 7). Countries at an early stage of implementation of fluorocarbon management regulations may be embarking on the process of upgrading existing ODS policies and regulations to include HFCs after the ratification of the Kigali Amendment.[29]

Countries at a medium stage of implementation are already establishing policies, projects, or regulations for the management of HFCs and ODS. For example, some countries may be putting in place policies or national legislation banning the venting of fluorocarbon substances. These policies may or may not be actively enforced given the challenges of LRM, such as lack of incentives or other economic barriers, leading to the venting of fluorocarbon substances from in-use equipment that is being serviced or managed for end-of-life, or other emissions.[30]

In the most advanced stages of implementation, Kigali Amendment measures are well-established and fluorocarbon management may be underpinned by strong efforts for collection, while reclamation activities may not yet be taking place. Such countries may be also establishing detailed regulations, guidelines, and standards for implementation of upstream, midstream, and downstream measures.[31]

The status of many Article 5 countries remains at early or medium stages of implementation of fluorocarbon management measures, and key solutions to meet the challenges of fluorocarbon management are needed, such as leveraging carbon finance.

An assessment evaluating the status of fluorocarbon management nationally is essential as a first step to understanding what strategic areas carbon markets may be able to support. Countries may begin this process by asking a series of key questions, identifying key issues with fluorocarbon management that they may want to address through carbon markets. Such questions may identify the underlying causes of non-optimal fluorocarbon management, such as the existence of stockpiles, instances of venting or lack of financing or enforcement, or the challenges in establishing waste collection networks (Figure 4).

To answer these questions, countries may need close cooperation between multiple government entities. For example, government officials from a National Ozone Unit may be already intimately knowledgeable with many of these questions and may be able to support such an assessment. Where the National Ozone Unit may be leading the effort, National Ozone Unit officials can coordinate efforts with a climate change or carbon markets government unit.

---

[29]   S. M. Taib, G. Nikoloski, and A. Arvanitidis. 2019. Cooling for the Future: The Case of HFC Alternatives in the Mediterranean Region. *Proceedings of the 2019 Heraklion Conference on Climate Change.* http://uest.ntua.gr/heraklion2019/proceedings/pdf/HERAKLION2019_MatTaib_etal.pdf.

[30]   Yale Carbon Containment Lab. 2023. Recovery and Destruction of Hydrofluorocarbon Refrigerant Gases in Article 5 Countries: Summary, Context, and Justification for Draft Methodology. White Paper. May 2023 Draft. https://carboncontainmentlab.org/documents/yale-cc-lab-hfc-methodology-white-paper-may-2023-draft.pdf.

[31]   A. Garg et al. 2023. *Activating Circular Economy for Sustainable Cooling: Global Best Practices on Lifecycle Refrigerant Management. Council on Energy, Environment and Water.* https://www.ceew.in/sites/default/files/global-best-practices-lifecycle-refrigerant-management-emissions.pdf.

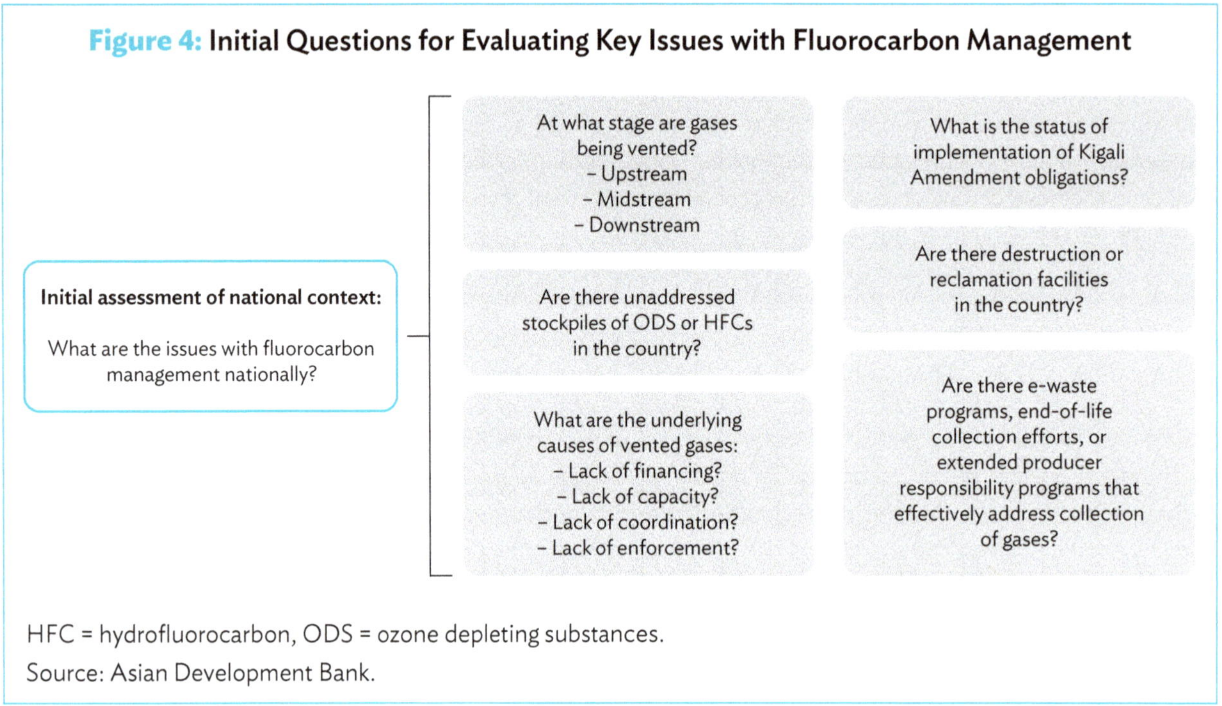
Figure 4: Initial Questions for Evaluating Key Issues with Fluorocarbon Management

HFC = hydrofluorocarbon, ODS = ozone depleting substances.
Source: Asian Development Bank.

In some cases, capacity building relevant to policymaking, designing projects, collecting data, and monitoring emissions may be needed for governments and for private or public sector stakeholders. Such capacity building can be a direct objective of operationalizing carbon markets for a comprehensive fluorocarbon life cycle management. At the public sector level, international climate finance or cooperation initiatives, including international development organizations, may support building national capacity, with the support of international experts or government-to-government cooperation and capacity building. At the private sector level, carbon projects may include in their design elements capacity building for local stakeholders, technicians, or other relevant players.

Understanding National Policies for Fluorocarbon Management

Countries must also understand the status of fluorocarbon management policies, laws, and regulations that are in place. Importantly, they must also understand the effectiveness of such policies at the present time in order to identify the issues that must be addressed.

For example, several countries have national laws and regulations that prevent the venting of gases during routine maintenance or at equipment end-of-life. However, these are often not enforced for reasons such as the complexity of monitoring millions of appliances containing fluorocarbons, lack of proper equipment required by technicians to capture the gas, and lack of reclaiming and destruction infrastructure in the country (footnote 30).

Other examples of e-waste policies or regulations, such as the Extended Producer Responsibility (EPR) or fluorocarbon waste disposal sites, may be in place or be a part of a government strategy for managing fluorocarbon gases. Such policies or regulations can be enhanced through the incentives that a carbon market brings, finding ways to meet financing, capacity needs, or other challenges that EPR and other LRM approaches may face. Technicians servicing existing equipment or disposing of end-of-life equipment, for example, may be provided a stronger incentive to properly dispose gases, through handing them over to a destruction facility or collection facility, if they receive a fee for doing so.

This fee can be generated through carbon market finance obtained through fluorocarbon carbon projects, thereby enhancing, promoting, or improving an already existing LRM policy. Overall, the leveraging of carbon market finance to promote such policies can be an effective way of providing finance where needed, as well as avenues for incentives, thereby helping achieve a more impactful fluorocarbon management. However, when considering the interaction between policies and carbon markets, policymakers and market participants must also closely consider the importance of designing regulations, policies, and projects with high integrity, keeping in mind the concept of additionality. In this context, "additional projects" are those that would not have occurred in the absence of carbon finance. The concept and importance of environmental integrity and additionality is expounded in section 2.3.5.

Some countries are also promoting a national market of reclaimed gases, that is, gases that have been collected from equipment, reconstituted to an appropriate quality level, and then made available for sale to newly put them into use. Creating such a market through direct policy interventions, however, may face some barriers. Reclaimed gases may not be sufficiently price-competitive against virgin gases, or consumers may not trust reclaimed gases because they are accustomed to purchasing virgin gases. A reclaimed gases market may be positively supported through carbon markets by putting incentives in place. For example, carbon market projects crediting the reclaiming of fluorocarbon gases would create a positive price to signal and reduce the cost of reclaimed gases against virgin gases, enhancing their price-competitiveness. This is another example of carbon finance supporting already existing policies in a strategic manner to achieve enhanced effectiveness or compliance.

In various national contexts, a combination of these policy interventions or regulations may already be in place to achieve national targets toward the Montreal Protocol and the Paris Agreement, or other national waste management targets. Overall, countries must understand the implementation and status of policies and regulations that have been described above nationally, in order to understand where carbon markets may have the most impact and how they may be leveraged effectively to support existing strategies and targets (Box 2). Once this step has been taken, countries are better prepared to identify strategic ways of leveraging carbon finance toward effective LRM.

Box 2

Classification of Hydrofluorocarbons as Hazardous Waste

Some regulatory and institutional frameworks that are not directly involved in carbon market mechanisms can be critical for implementing hydrofluorocarbon (HFC) destruction projects. There are diverse issues depending on the domestic framework of each Asian Development Bank developing member country (DMC). This publication introduces a case study from TA 6730: Promoting Life Cycle Management of Fluorocarbons on hazardous waste regulation which can be common across many DMCs.

Collected HFCs may be categorized as hazardous waste under domestic laws and regulations. Many DMCs follow the Basel Convention's list of hazardous wastes, which includes fluorocarbons and fluorocarbon-containing equipment. Hazardous waste must be transported by a licensed organization, wherein refrigeration mechanics are rarely a part of. According to the Technology and Economic Assessment Panel (TEAP) 2024 Task Force Report on Life Cycle Refrigerant Management, there are two options for recovered fluorocarbon to be exempted from being classified as hazardous waste: (i) materials recovered for the purpose of recycling or reclamation are not waste and are simply reprocessed original substances with intent of reuse; and (ii) recovered fluorocarbons should be classified as non-hazardous as far as it has no direct local environmental and human health impacts in comparison to the original product.

continued on next page

Box 2 *continued*

Viet Nam is one of the pioneering DMCs in fluorocarbon recovery, reclamation, and destruction. The potential of recovered fluorocarbon being classified as hazardous waste was noticed during the establishment of the country's first commercial reclamation operation. Refrigerants and equipment are listed as hazardous waste in Annex III of the MONRE Circular 02/2022/TT-BTNMT. Viet Nam is moving forward with option 1 above. Until collected fluorocarbon is tested and classified as inapplicable for reclamation, it can be simply regarded as fluorocarbon. From the testing site to destruction site, the fluorocarbon would be regarded as hazardous waste. This approach is viable only if a reclamation is implemented or is being planned.

The Philippines has Department of Environment and Natural Resources Administrative Order 2021-31 on Chemical Control Order for HFCs, which requires fluorocarbon importers, dealers, resellers, retailers, and service providers to hold a Hazardous Waste Generator Certificate at registration. This demanding requirement can be one of the barriers to registration as a fluorocarbon recovery entity. There is only one entity licensed so far.

Even if the recovered fluorocarbons deteriorate and are inappropriate for reclamation, it is less likely to contain other hazardous components except for a traceable amount of hydrogen halides and machine oil. In this regard, option 2 may be justified for fluorocarbons. Countries that do not have reclamation facilities and do not plan to build one may need to take this option or provide any other justification for fluorocarbon transportation to implement a destruction project.

Source: United Nations Environment Programme. TEAP May 2024: Decision XXXV/11 Task Force Report on Life Cycle Refrigerant Management (Volume 3). https://ozone.unep.org/node/15193.

2.1.2 Identifying Sources of Fluorocarbon Gases

Countries looking to strategically leverage carbon markets must also understand the potential emission sources of fluorocarbons in their jurisdictions. These may include a variety of sources and may have different implications. Figure 5 shows some potential sources of fluorocarbon emissions existing in countries.

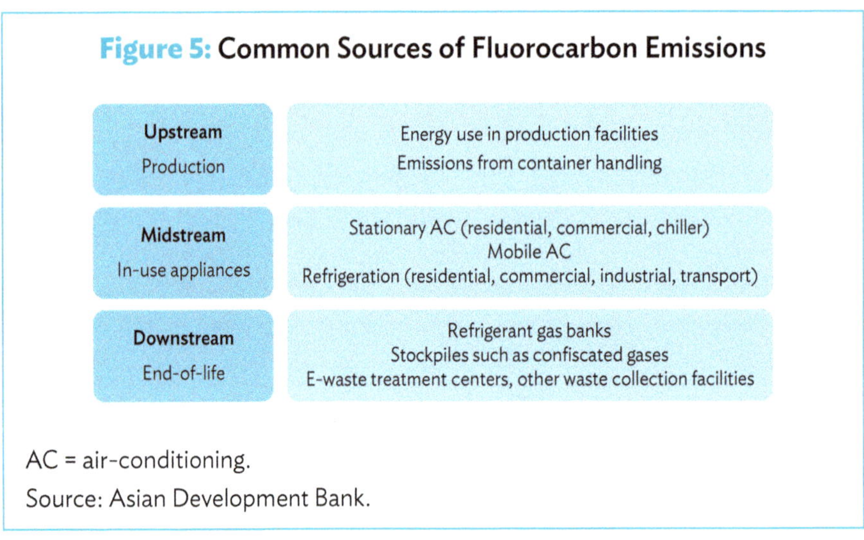

Figure 5: Common Sources of Fluorocarbon Emissions

AC = air-conditioning.
Source: Asian Development Bank.

These sources may represent project scenarios where the mitigation of emissions arising from the source can be credited and therefore receive carbon finance. Countries may also wish to reference these sources of fluorocarbon gases against Figure 1, identifying the potential solutions that may be enabled by carbon finance.

2.1.3 National Readiness to Participate in International Carbon Markets

Defining readiness for international carbon markets depends on the context being addressed: Are Article 6.2 carbon markets being prioritized, Article 6.4, or voluntary carbon market (VCM)? Countries may also want to ask themselves what types of strategic objectives are they aiming to address. These may determine which readiness steps, described in more detail below, they may need to take.

In understanding their national readiness to participate in Article 6, countries must make several arrangements, outlined through the UNFCCC, to meet participation requirements. This topic is covered briefly here but is discussed in more detail in ADB's knowledge product *National Strategies for Carbon Markets Under the Paris Agreement*.[32]

On the side of the VCM, many countries, both in Asia and other regions, host a range of VCM activities nationally; however, country governments may not be overseeing closely the types of projects taking place. Generally, many countries already have active voluntary carbon market projects, and given the voluntary nature of the VCM, no direct participation "readiness" is needed. However, in order to strategically enable fluorocarbon management projects to take place, countries may also wish to identify ways to build capacity generally and promote and leverage VCM projects toward achieving fluorocarbon management objectives. This may involve regulatory change, promotion efforts, or awareness programs.

In addition to participation requirements for Article 6 and general VCM participation, fluorocarbon management projects rely on specific infrastructure and facilities as a means of managing gases. For example, projects destroying fluorocarbons will rely on destruction facilities. Readiness for developing a pipeline of fluorocarbon projects will involve knowledge of the destruction facilities that exist, if any, and whether they meet the key requirements in project methodologies. Projects themselves, as part of design and implementation, may directly involve the construction or modification of a facility to destroy or reclaim fluorocarbons.

Readiness for Article 6 Carbon Markets

The Paris Agreement's Article 6 alludes to several country participation requirements to engage with both Article 6.2 and Article 6.4. The requirements for participation in Article 6 of the Paris Agreement for countries are summarized in Figure 6.

Global best practices on how to meet these requirements are emerging. Of the requirements listed in Figure 6, establishing *arrangements for authorizing, tracking, and reporting the use of ITMOs toward the achievement of NDCs* represents a significant task ahead of countries looking to participate in Article 6.2.

---

[32] Asian Development Bank. 2023. *National Strategies for Carbon Markets Under the Paris Agreement: Making Informed Policy Choices.* https://www.adb.org/sites/default/files/publication/928596/national-strategies-carbon-markets-paris-agreement.pdf.

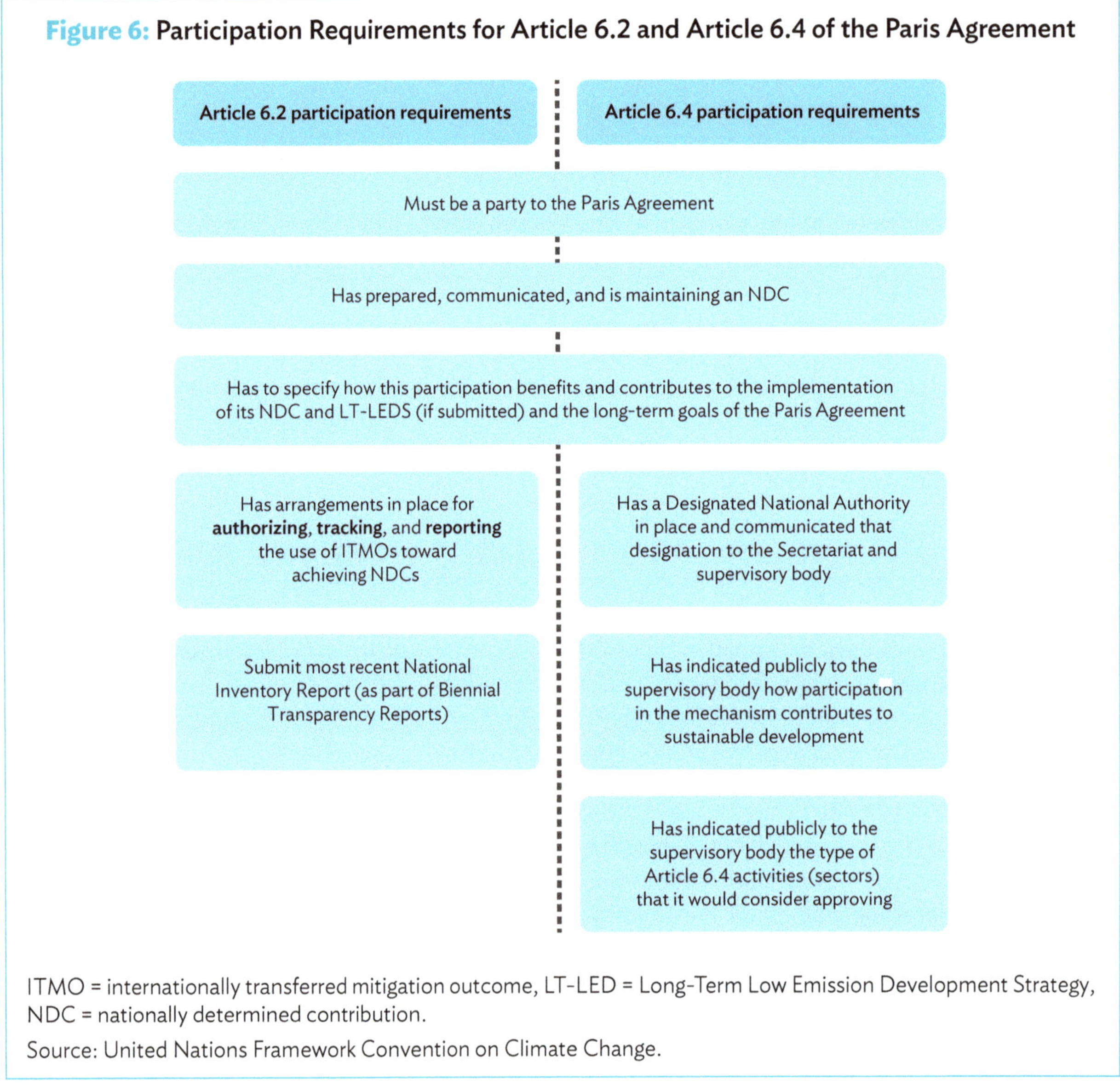

Figure 6: Participation Requirements for Article 6.2 and Article 6.4 of the Paris Agreement

ITMO = internationally transferred mitigation outcome, LT-LED = Long-Term Low Emission Development Strategy, NDC = nationally determined contribution.
Source: United Nations Framework Convention on Climate Change.

Many front-runners are now developing frameworks and tracking arrangements for ITMOs to meet these requirements. A comprehensive Article 6 participation framework requires arrangements for authorizing, tracking, and reporting (Figure 7).

Countries might also benefit from reviewing the structure of institutional arrangements to understand any limitations or amendments that may be needed. In many cases, a dedicated National Ozone Unit exists to manage the implementation of the Kigali Amendment and Montreal Protocol.

Furthermore, countries will have already begun their Article 6 journey, and have completed an Article 6 framework, or may be in the process of developing one at the time that strategically leveraging carbon markets toward fluorocarbon management is being considered. Overall, it is important to coordinate between government units' progress in developing such frameworks and regulations in order to have an up-to-date understanding of readiness status for Article 6.

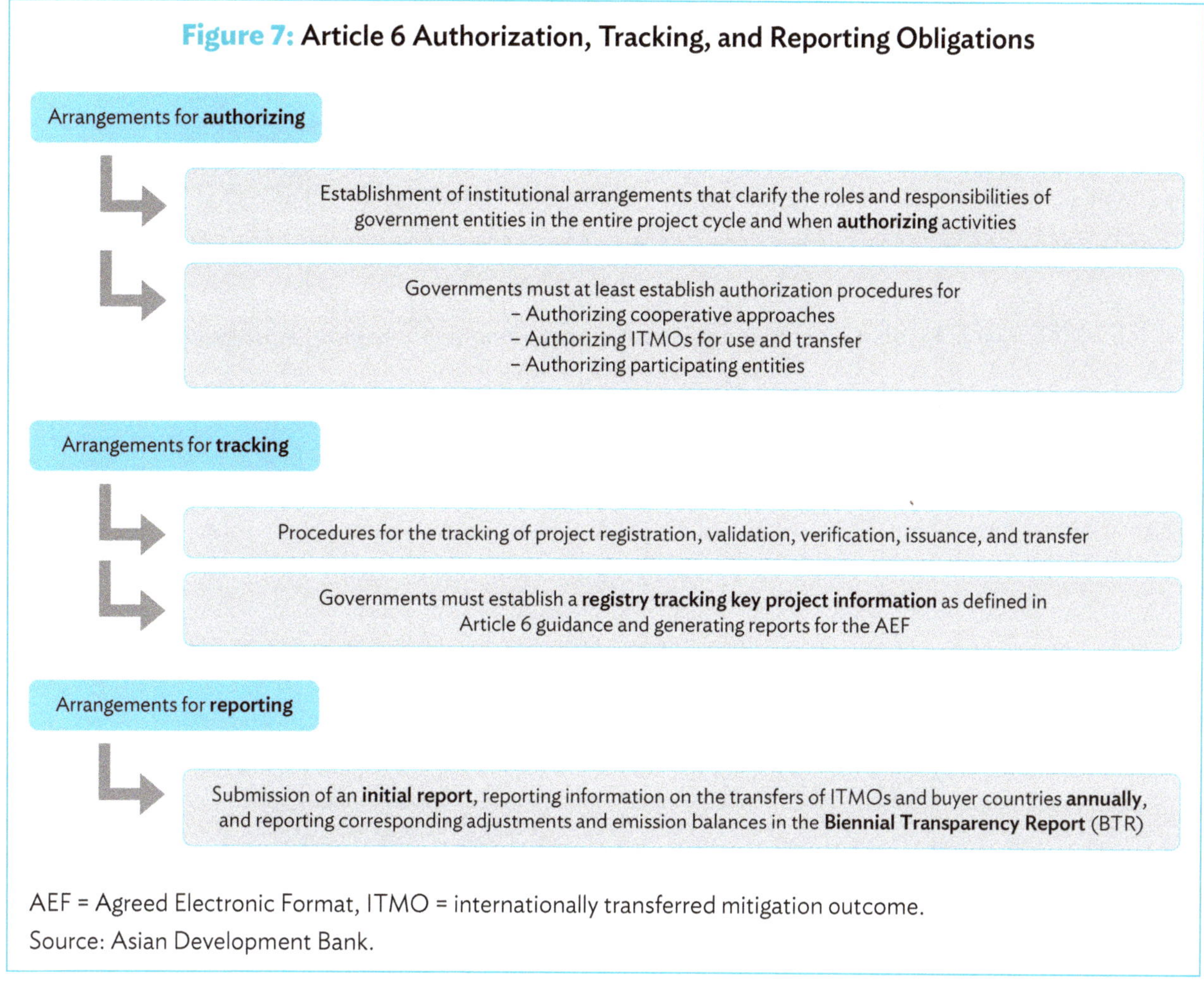

AEF = Agreed Electronic Format, ITMO = internationally transferred mitigation outcome.
Source: Asian Development Bank.

Institutional Arrangements for Fluorocarbon Management Under Article 6

Institutional arrangements are defined as the governance entities within a country, such as ministries, ministerial agencies, government organizations, or those that may have government mandates to regulate, oversee, or pursue Article 6 cooperation on behalf of a country.

Countries exploring participation in Article 6 also must consider their current institutional arrangements for managing carbon market activities. These may include, for example, those established under the Clean Development Mechanism (CDM). New requirements under Article 6 of the Paris Agreement may require some institutional arrangement review, which may also be influenced by the types of mitigation projects they may want to pursue, such as fluorocarbon management.

As part of their participation in Article 6 of the Paris Agreement, countries are revising or establishing new institutional arrangements, often based on existing examples or previous experience in the CDM. Institutional arrangements are of key importance for the efficient, orderly, and transparent management of participation in Article 6 and for supporting various actors, whether it be on the public or private side, and in fulfilling Article 6 requirements and effectively pursuing projects. Specifically, institutional arrangements, which determine the responsibilities and mandate of entities participating in Article 6, play a role in meeting Article 6 requirements.

Currently, many countries that are already participating in Article 6 have various institutional setups, working with existing legal mandates, or establishing new ones supporting participation requirements, administrative tasks, and overall coordination. These institutional setups are also helping manage cooperative approaches efficiently. An example of institutional arrangements for managing Article 6 approaches nationally is shown in Figure 8. Relative to national context, these may be proposed differently, depending on the legal structures and government mandates of a country.

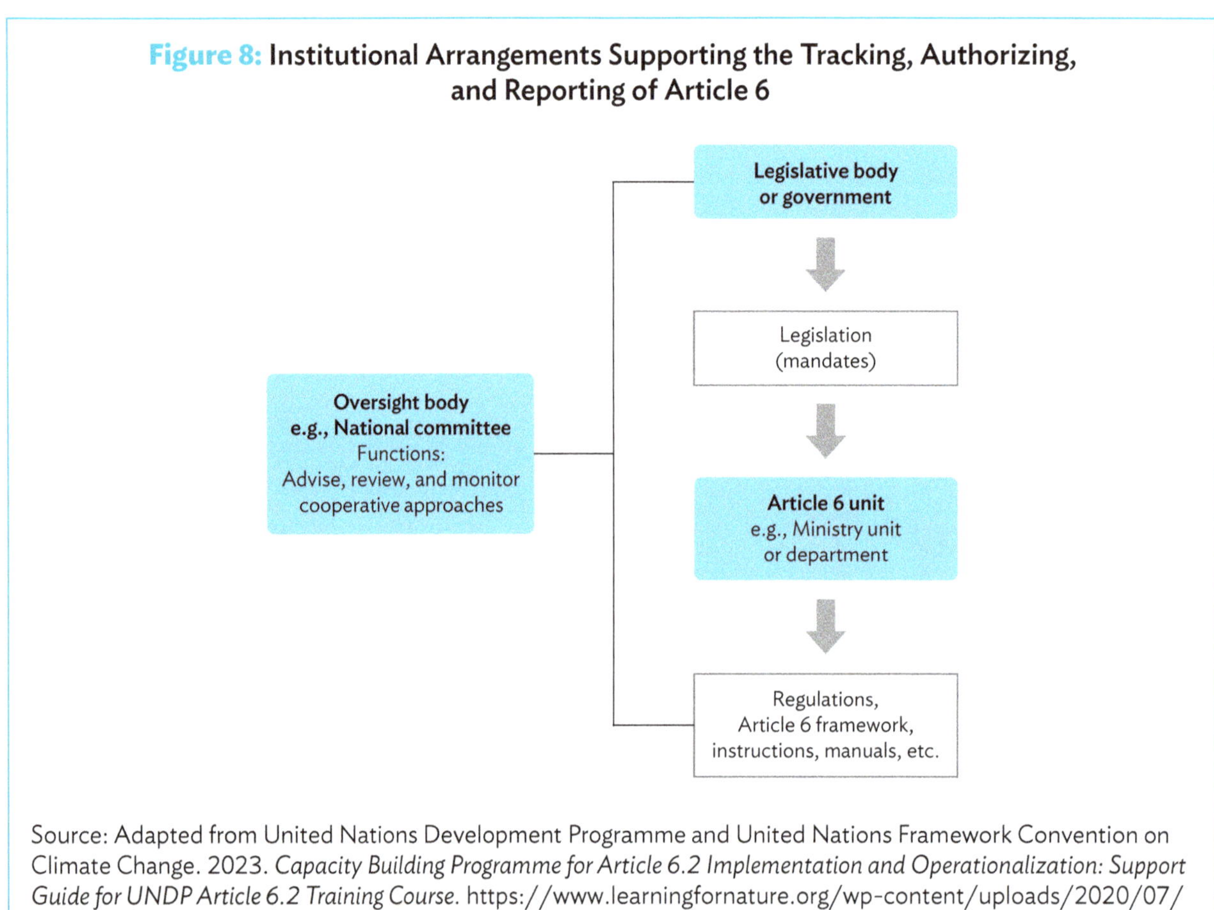

Figure 8: Institutional Arrangements Supporting the Tracking, Authorizing, and Reporting of Article 6

Source: Adapted from United Nations Development Programme and United Nations Framework Convention on Climate Change. 2023. *Capacity Building Programme for Article 6.2 Implementation and Operationalization: Support Guide for UNDP Article 6.2 Training Course.* https://www.learningfornature.org/wp-content/uploads/2020/07/Support_Guide_UNDP_UNFCCC_23.01.2023-compressed.pdf.

Within the context of fluorocarbon management, institutional arrangements are of particular importance, given the already existing experience of managing fluorocarbons as part of the Montreal Protocol and the number of stakeholders involved. There is often strong knowledge within National Ozone Units or other environmental units that manage targets under the Montreal Protocol, particularly in terms of the national context; monitoring, reporting, and verification (MRV) setup; stakeholders; and other matters specifically relating to fluorocarbon management.

However, the government units or entities tasked with overseeing, regulating, and governing Article 6 and carbon markets, including decision-makers in authorization of projects, are often not the same as the ones overseeing fluorocarbon management and activities nationally. Often, a climate or national carbon market government unit will not have direct insights into fluorocarbon management or tasks in the achievement of the Kigali Amendment and Montreal Protocol. This often may be a barrier to coordination between units.

In order to address this challenge, a National Ozone Unit or other similar unit can support the development of carbon market activities aiming to transform fluorocarbon life cycle management through carbon finance. As mentioned previously, such a unit often has good technical knowledge and expertise and can provide accurate data, as well as support for developing baselines and reviewing projects (e.g., review of project ideas and detailed project studies and documentation). Furthermore, it may be useful to have National Ozone Unit experts supporting the evaluation of projects and documentation received to ensure good project design, evaluate technical aspects, and review environmental integrity risks.

While institutional arrangements may be similar to the ones proposed above in specific national contexts, they may also be vastly different due to different mandates and legal structures. In such cases, countries may benefit from identifying the gaps and needs to be addressed to fully identify the barriers, opportunities, and challenges that technical expertise, coming from a National Ozone Unit or other sources of fluorocarbon sector expertise, can tackle. In all such cases, technical expertise and on-the-ground, day-to-day experience in the life cycle management of fluorocarbons can effectively help inform strategic ways to achieve positive mitigation outcomes and the types of projects that may be pursued, while serving to fulfill Article 6 requirements.

In Figure 9, two different points are proposed to include technical expertise or support from a National Ozone Unit. On one hand, a national oversight body, consisting of multiple government sectors and ministries, may have inputs from a National Ozone Unit or from another unit managing the implementation of the Montreal Protocol, such as in the form of an expert advice. Similarly, institutional arrangements currently being put in place by a range of Article 6 participant countries include an Article 6 unit under a relevant ministry or technical committee. This unit or technical committee may be supported by technical expertise from the National Ozone Unit or from the unit responsible for Montreal Protocol implementation.

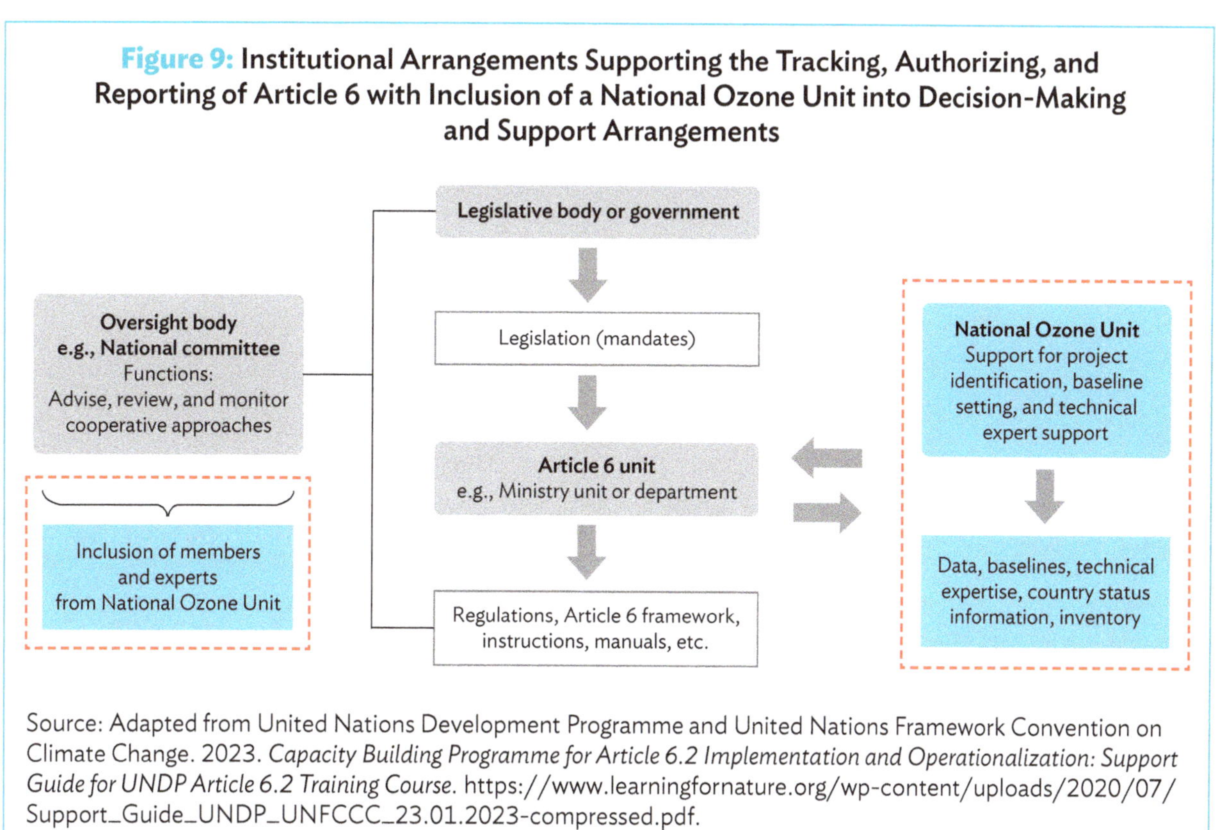

Figure 9: **Institutional Arrangements Supporting the Tracking, Authorizing, and Reporting of Article 6 with Inclusion of a National Ozone Unit into Decision-Making and Support Arrangements**

Source: Adapted from United Nations Development Programme and United Nations Framework Convention on Climate Change. 2023. *Capacity Building Programme for Article 6.2 Implementation and Operationalization: Support Guide for UNDP Article 6.2 Training Course.* https://www.learningfornature.org/wp-content/uploads/2020/07/Support_Guide_UNDP_UNFCCC_23.01.2023-compressed.pdf.

While these arrangements may vary across countries, government units may benefit from technical expertise and technical information that may currently be present in a National Ozone Unit.

2.1.4 Screening National Climate Policies: Nationally Determined Contribution and National Inventory

Article 6 carbon markets may be leveraged to support achievement of NDCs through collaboration between countries. Given their role, Article 6 carbon markets are tightly linked with NDCs, which are in turn closely linked with national inventories.

NDCs and national inventories are the two most relevant national climate instruments impacting the development of fluorocarbon projects through Article 6. NDCs represent central policy documents detailing targets and ambition toward tackling climate change and achieving the objectives of the Paris Agreement, while national inventories are the primary accounting tool countries use to track national emissions, including fluorocarbons, specifically HFCs.

Screening both the NDC and the national inventory for readiness are an important step to determining Article 6 participation readiness, particularly to enable transparent corresponding adjustments as part of the authorization of ITMOs. Two key criteria are needed:

(i) The NDC **should** include both the sector (Industrial Processes and Product Use) and the gas type (HFC) within its scope and coverage.

(ii) The national inventory **must** report on the gas type (HFC) when transacting ITMOs arising from HFC mitigation, based on rules under the Enhanced Transparency Framework.[33]

These two criteria are either highly recommended or essential for transacting ITMOs from HFCs under Article 6 of the Paris Agreement and thus must be considered when reviewing national context and readiness. The accounting and reporting rationale, including the requirements for corresponding adjustments to correctly account for emission reductions, are described in more detail in Chapter 2.4— Accounting and Reporting for Fluorocarbon Projects within the Montreal Protocol and Paris Agreement.

Countries may also want to evaluate the type of target they have included or may include: Is it a conditional or an unconditional target? Broadly, countries choosing to pursue HFC projects through Article 6 may prefer to include HFCs in a conditional target, where international support is required for its achievement. This may transparently indicate to international buyers that Article 6 may be a means of collaborating to achieve the conditional target.

2.2 Strategic Fluorocarbon Management Through Carbon Markets

The strategic leveraging of finance tools, such as carbon markets, can enable effective fluorocarbon management approaches. To date, countries have already leveraged carbon markets to create positive, cost-effective outcomes for their fluorocarbon management efforts nationally.

---

[33] UNFCCC, CMA. 2016. Decision 18/CMA.1. Para 48. Modalities, Procedures and Guidelines for the Transparency Framework for Action and Support Referred to in Article 13 of the Paris Agreement. https://unfccc.int/resource/tet/0/00mpg.pdf.

One notable example of this is the reclaimed market in the US, where strong national regulations all come together to drive the achievement of Kigali Amendment targets, emission crediting standards for fluorocarbons, and a reclaimed market.[34] This is an example of strategically utilizing carbon markets to fund emission reductions in a sector that needs innovative financial solutions, where otherwise the cost of abatement typically falls on producers and operators.

This section identifies ways in which countries may go about identifying and setting strategic objectives that can be achieved by leveraging carbon finance for fluorocarbon management. Examples of strategic objectives are also included.

2.2.1 Identifying and Setting Strategic Objectives

Countries may begin to consider the fluorocarbon management objectives that they may want to achieve, that is, the specific areas that they may want to target to achieve emission reductions. One way of doing so is to look at the national context and emission baseline issues that may have been identified as part of initial investigations or national challenges in fluorocarbon management generally (Figure 10).

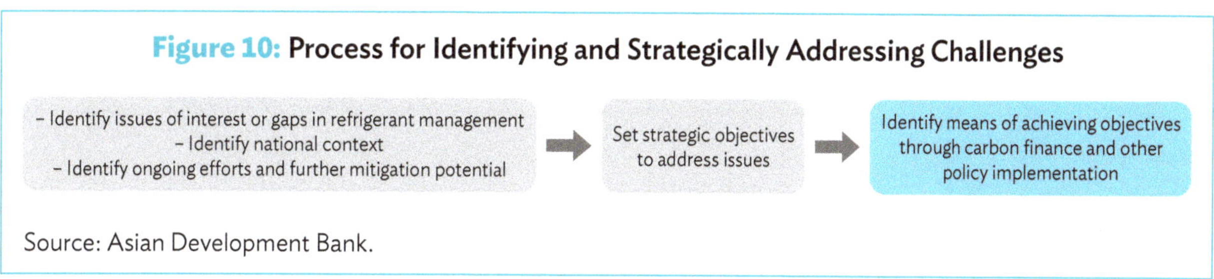

Source: Asian Development Bank.

The national context of existing capacity, facilities, and experience with fluorocarbon management is also significant, particularly when it comes to destruction facilities, reclamation facilities, existing crediting projects, national project developers, and ongoing efforts in fluorocarbon management and the achievement of the Montreal Protocol and Paris Agreement targets.

Once issues have been identified, policymakers can make informed decisions and choices of what types of carbon crediting may be most effective in addressing such issues.

2.2.2 Strategic Objectives That Can Be Supported by International Carbon Markets

The types of strategic objectives are highly dependent on national priorities, policies, and circumstances surrounding the achievement of national objectives, as well as the means and expertise available.

However, countries do face common challenges when addressing fluorocarbon management. Figure 11 lists examples of common national challenges addressing fluorocarbon management at upstream, midstream, and downstream stages, as well as the potential objectives that may be derived or identified based on these challenges.

---

[34] A. Garg et al. 2023. Activating Circular Economy for Sustainable Cooling: Global Best Practice on Lifecycle Refrigerant Management. Council on Energy, Environment and Water. https://www.ceew.in/sites/default/files/global-best-practices-lifecycle-refrigerant-management-emissions.pdf.

Figure 11: National Refrigerant Management Challenges and Strategic Objectives

	Examples of national challenges to be addressed	Examples of strategic objectives
Upstream Production	• Meeting Kigali Amendment targets on the production of refrigerants • No nationally reclaimed or recycled refrigerant market • Reclaimed/recycled refrigerants are not price competitive with virgin gases • No reclamation facilities existing nationally	• Leveraging carbon finance to drive down cost or accelerate reduction of production of virgin HFCs • Create a national reclaimed and recycled refrigerant market • Make existing reclaimed or recycled refrigerants more price competitive with virgin production • Catalyze the construction of a reclamation facility
Midstream In-use appliances	• In-use appliances are energy inefficient or use outdated refrigerants • When in-use appliances are serviced, refrigerants are often vented • Lack of compliance with existing or new regulations due to lack of incentives	• Promote or drive down cost of replacement of inefficient or non-environment-friendly in-use refrigerant appliances • Promote energy-effcient appliances • Prevent the intentional venting of refrigerant gases when in-use appliances are serviced • Provide incentives for compliance with existing or new regulations for refrigerant management
Downstream End-of-life	• Many existing stockpiles are at risk of leaking refrigerants • No organized collection or management efforts of end-of-life equipment are in place • No facility exists for the destruction of end-of-life refrigerant gases • Lack of compliance with existing or new regulations due to lack of incentives • Lack of cost-effective refrigerant collection networks nationally	• Destroy or address existing stockpiles • Promote or incentivize cost-effective programs for collection or management of end-of-life equipment • Catalyze the construction of a refrigerant destruction facility • Promote or incentivize compliance with existing or new regulations • Foster collection networks by providing incentives to implementors

HFC = hydrofluorocarbon.
Source: Asian Development Bank.

Policymakers may then draw direct links between a challenge they are currently facing in the area of refrigerant management and an objective they may choose to pursue as part of a wider strategy. Such objectives may already be a part of a national Long-Term Low Emission Development Strategy, an NDC, a strategy to achieve Kigali Amendment and Montreal Protocol targets, or other strategies. Alternatively, such strategies may contain targets or objectives at a higher level.

Generally, a wide range of challenges share the common requirement of finance or financial incentives. Many countries have already identified areas in need of support, capacity, and finance, such as those identified here. Carbon markets can be efficient instruments to provide this finance, and policymakers may choose to achieve such targets or objectives by leveraging carbon finance.

2.3 Creating a Refrigerant and Fluorocarbon Project Pipeline

Once actively engaging with carbon markets, governments may face the challenge of identifying, evaluating, authorizing, and regulating the projects they receive. They may also receive an abundance of project ideas and may be challenged to review many of these. Alternatively, at early stages, they may wish to receive more project ideas and manage a project pipeline more efficiently.

Initially, awareness and knowledge must be built on the available project types, technologies, and means of addressing fluorocarbon emissions.

This section provides technical advice on these matters and toward creating a high-quality refrigerant project pipeline from a national government perspective. Initially, a description of how projects may be identified for the national government is provided, beginning with the means of addressing fluorocarbon emissions through technologies and project types.

Further technical issues are described from a host country and project developer perspective to operationalizing projects, including project eligibility, viability, finding applicable methodologies, and identifying buyers.

Given the importance of maintaining high integrity, a description of how the concept applies to carbon projects through both the voluntary carbon markets and Article 6 is also provided.

Lastly, innovative ways to leverage carbon finance are introduced, specifically the concept of a policy crediting approach.

2.3.1 Identifying Means of Addressing Fluorocarbon Life Cycle Emissions

A range of technologies, measures, and policies have been designed and implemented to address fluorocarbon emissions at various stages of their life cycle, from production to end-of-life. Part of the process of establishing a refrigerant project pipeline involves understanding the types of technologies and projects that may be eligible to receive carbon finance.

The *Report of the Technology and Economic Assessment Panel*, published in May 2024, gives a very comprehensive overview of LRM measures, including challenges, opportunities, and strategies, and aims to provide stakeholders with the necessary knowledge for addressing fluorocarbon management effectively (footnote 7). A detailed summary of the available technologies for leakage prevention, recovery, recycling, reclamation, and destruction, and an assessment of their accessibility in parties operating under paragraph 1 of Article 5 of the Montreal Protocol are provided.

Another study by the Climate and Ozone Protection Alliance (COPA) explores the solutions available for addressing ODS/HFC banks, including a detailed summary on the conditions, challenges, available technologies, and current state of ODS/HFC reclamation and destruction practices in Article 5 countries.[35]

Once countries have identified sources of emissions from fluorocarbon applications, they may wish to identify options available within the country for addressing them within the framework of carbon markets.

---

[35] Coalition of Organizations for the Protection of the Atmosphere. 2024. *Destruction and Reclamation of Ozone-Depleting Substances and HFCs.* https://www.copalliance.org/imglib/publications/COPA_Publication_ODS_HFC_%20 Destruction_Reclamation.pdf.

This currently includes the following five options: (i) reclamation, (ii) destruction, (iii) exporting gases, (iv) transition to green cooling, and (v) leak detection and prevention.

(i) Reclamation

The Technology and Economic Assessment Panel (TEAP) defines reclamation as the reprocessing and upgrading of a recovered substance through various means, such as filtering, drying, distillation, and chemical treatment, done with the purpose of restoring the substance to a specified performance standard (footnote 7).

This is done in licensed facilities. A commonly referred standard is the Standard 700-2016 from the Air-Conditioning Heating and Refrigeration Institute of the US, which stipulates that fluorocarbons need to reach 99.5% purity level to be considered reclaimed gases. Recycling, on the other hand, is a cleaning process that is usually done on-site, and the uncertified product is returned to the same equipment. Where possible, recycling or reclamation of HCFCs and HFCs may be preferable over destruction for environmental reasons and to achieve a circular economy (this is not relevant for CFCs, since their use is already prohibited due to the Montreal Protocol phaseout).

Besides establishing whether there are reclamation facilities within the country that can reprocess fluorocarbons to the required purity levels, the following main challenges to reclamation in Article 5 countries were identified in the COPA study:

- Ensuring a steady flow of fluorocarbons for reclamation
- The regulatory framework of the country
- Commercialization of the reclaimed fluorocarbons

The regulatory framework of the country is relevant in this context because obtaining permits to operate and handle hazardous substances like fluorocarbons is often a challenge since reclamation is not a regulated activity in many countries or is very new for the authorities. Also, the price of virgin HFCs, which has not increased as expected in most Article 5 countries, coupled with the lack of knowledge about these gases making consumers reluctant to switch to reclaimed gases, makes it difficult to establish price-competitive reclaimed gas markets.

Overall, the costs of reclamation activities will vary widely depending on a range of factors, such as the composition of gases, staffing costs, and the costs of equipment. TEAP identifies several potential costs associated with the setup of such reclamation activity types.[36]

Activities with a potential revenue stream, such as the reclaiming and resale of refrigerants, may generate revenues from the commercial operation of the underlying activity, as well as from carbon finance arising from the generation of carbon credits. This feature will also impact the final price for the carbon credits generated by the activity. In such cases, financial additionality is an important environmental integrity item that must be examined on a case-by-case basis.

---

[36]   United Nations Environment Programme. 2024. *Montreal Protocol on Substances that Deplete the Ozone Layer: Report of the Technology and Economic Assessment Panel. Decision XXXV/11 Task Force Report on Life Cycle Refrigerant Management.* (See Table 7.2). https://ozone.unep.org/system/files/documents/TEAP-May2024-DecXXXV-11-TF-Report.pdf.

(ii) Destruction

Although destruction is globally a more common practice than reclamation, due to a lack of appropriate regulatory frameworks, financial means, and infrastructure, the destruction of ODS and HFCs still faces many barriers, especially in Article 5 countries. Therefore, destruction projects may be highly additional, based on national context, and a viable option as part of the strategic life cycle management of fluorocarbons. Countries wishing to leverage carbon finance for promoting or enhancing destruction levels of end-of-life or other sources of fluorocarbons nationally must consider the national context and how such projects can achieve strategic objectives. In addition to the assessment of the regulatory and legal frameworks, Figure 12 contains the considerations on the available destruction technology in the country that needs to be reviewed.

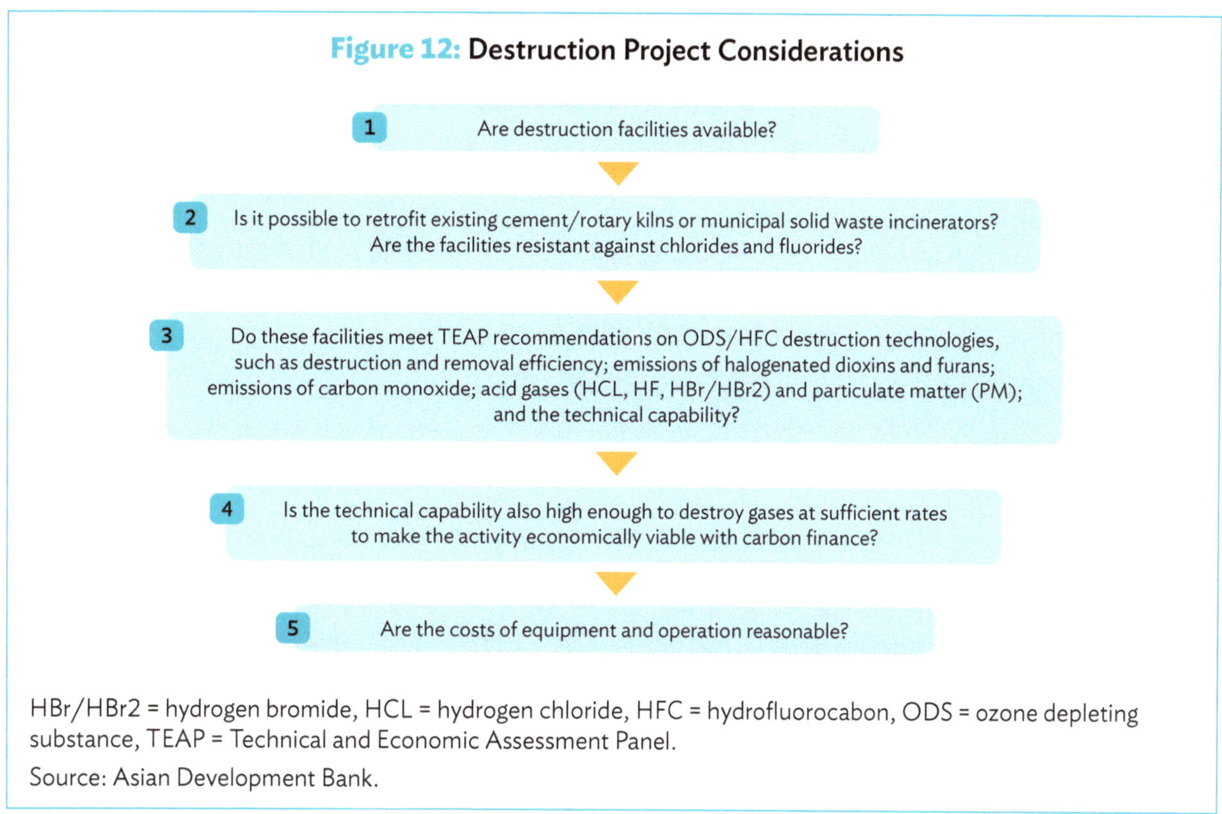

Figure 12: Destruction Project Considerations

HBr/HBr2 = hydrogen bromide, HCL = hydrogen chloride, HFC = hydrofluorocabon, ODS = ozone depleting substance, TEAP = Technical and Economic Assessment Panel.

Source: Asian Development Bank.

The costs of conducting destruction activities will vary depending on a range of national circumstances and contexts. For destruction activities, different sources identify destruction costs, that is, costs incurred only in the chemical decomposition of fluorocarbons, between \$2 and \$3 per kilogram for the lowest estimates (footnote 7), assuming economies of scale and optimistic contractual arrangements, to \$6 per kilogram or more, in cement kiln applications, which are often the cheapest technological solution.[37] This does not include the costs of construction or retrofitting of a facility, which TEAP estimates to range between \$50,000 and \$100,000. These costs can vary greatly depending on national circumstances and the project context.

---

[37]   Coalition of Organizations for the Protection of the Atmosphere. 2024. *Destruction and Reclamation of Ozone-Depleting Substances and HFCs*. https://www.copalliance.org/imglib/publications/COPA_Publication_ODS_HFC_%20 Destruction_Reclamation.pdf.

As with all carbon market projects, it is important to note that such price estimations are not indicative of final carbon price. A full carbon price for a credit arising from fluorocarbon mitigation will include other important costs incurred in the development of such an activity. These may include training costs of technicians, capital expenditures for equipment such as cylinders, permitting costs, collection costs, maintenance, chain of custody, and other items. In carbon market projects receiving no inherent revenue from the underlying activity, such as in the case of the destruction of fluorocarbons, the carbon credit price is expected to cover all of the above costs.

Within the context of carbon market projects in fluorocarbons, TEAP requirements are often referenced as criteria for eligibility of a fluorocarbon destruction facility. TEAP provides requirements on the eligibility of a fluorocarbon destruction facility. However, there is no specific approval by TEAP of a facility. Instead, TEAP works to approve technology types, not facilities, and has several screening criteria for technologies solely for the purpose of destroying fluorocarbons in compliance with the Montreal Protocol, rather than carbon markets. Therefore, there is no direct approval by TEAP or a "TEAP-approved facility" within a carbon market context, and there is no specific body issuing a TEAP certification for facilities.

The Montreal Protocol additionally identifies the types of technologies eligible for destruction of substances under the treaty, which are approved by the signatory countries.[38] The most recent list of destruction technologies approved by parties to the Montreal Protocol is contained in Annex II to the 30th Meeting of the Parties to the Montreal Protocol under Decision XXX/6, as amended by Decision XXXV/6.[39] While this list is intended primarily for destruction at production of fluorocarbons, the destruction of recovered fluorocarbons may be carried out with other technologies.[40]

The approved technologies can be grouped into three general categories:

- Thermal oxidation
- Plasma technologies
- Chemical transformation technologies which are generally applicable to processes intended to recover chemicals that are not controlled substances for reuse within a production or manufacturing process

According to TEAP, for practical purposes the technologies that are available and realistically accessible for the destruction of end-of-life fluorocarbons will be within the categories of thermal oxidation, plasma arc, and cement kiln (footnote 7).

These TEAP requirements are translated into an eligibility check which is often used within carbon market contexts and often included within carbon crediting methodologies under various standards. This is considered best practice within carbon market projects in fluorocarbon applications. A common eligibility check consists of the following:

- The facility meets destruction and removal efficiencies requirements outlined by TEAP (verifiable by validation and verification bodies [VVBs] which are independent third-party auditing firms approved or accredited by standards or registries).[41]

---

38   UNIDO. 2018. *Destruction Technologies for Ozone-Depleting Substances under the Montreal Protocol.* https://www.waste-management.org/en/destruction_technologies/montreal_protocol/.

39   United Nations Environment Programme. 2024. *Destruction Procedures under the Montreal Protocol.* https://ozone.unep.org/treaties/montreal-protocol/destruction-procedures.

40   United Nations Environment Programme. 2024. *Montreal Protocol: Article 2J – Hydrofluorocarbons.* https://ozone.unep.org/treaties/montreal-protocol/articles/article-2j-hydrofluorocarbons.

41   Technology and Economic Assessment Panel. 2018. *TEAP Report: April 2018.* https://ozone.unep.org/sites/default/files/2019-04/TEAP-DecXXIX4-TF-Report-April2018.pdf.

■ The facility meets all national permitting requirements for air pollution (verifiable by a VVB, but contingent on regulatory compliance).

While based on these requirements, no direct government sign-off for the facility is required in voluntary or Article 6 carbon market projects, compliance with national regulations is necessary, and further requirements may be included in a project methodology. In addition, governments may choose to mandate some of these checks. Carbon market project developers experienced in the design and development of fluorocarbon application projects may be well experienced in these requirements and can therefore recall existing capacity and knowledge.

(iii) Exporting fluorocarbon gases

In some cases, such as when there are not enough fluorocarbons available for reclamation or destruction, it may be more economically attractive to export these gases to another country where facilities are available. Detailed analyses should be conducted to compare the economics of setting up destruction/reclamation facilities to exporting the gases. In such cases, countries may wish to identify a strategy for assessing such projects and any links to relevant national legislation.

Such exports will be subject to the rules of the Basel Convention on the Control of Transboundary Movements of Hazardous Wastes and their Disposal (Basel Convention).[42] The Basel Convention is an international agreement that regulates the transboundary movements of hazardous wastes and other wastes with the aim of protecting human health and the environment against the adverse effects resulting from the generation, transboundary movement, and management of these wastes. Specifically, one of its main goals is to ensure that hazardous waste is not being moved from developed countries to less developed ones unless there is prior informed consent, and the receiving country has operational facilities adopting international best practices to manage the wastes.

Any mitigation activities that involve the movement of controlled fluorocarbons between countries will need to evaluate the impacts of the Basel Convention on the project activity. In addition, projects developed under the Article 6 crediting mechanisms will have to assess how the ITMOs generated will be shared between the source country and the country where the emission reduction occurs. This sharing needs to be arranged and negotiated between the Parties involved.

(iv) Transition to green cooling

Carbon markets could play a key role in transforming the cooling sector and enabling the transition to climate-friendly and energy-efficient cooling equipment. The project type involves applications where natural refrigerants, such as propane, isobutane, ammonia, or carbon dioxide (CO_2), are used in combination with more energy-efficient equipment.[43] These projects can also result in significant emission reductions as well as a range of other co-benefits, including employment creation and technology transfer.

Historically, the uptake of these project types in the carbon markets has been limited. However, Article 6, given its project and policy crediting mechanisms, could be instrumental in financing these transitions. One large-scale green cooling project through Article 6 is currently being implemented in Southern Africa by the Deutsche Gesellschaft für Internationale Zusammenarbeit (GIZ) and GFA Consulting Group; it aims to

---

[42] UNEP. 2011. Basel Convention on the Control of Transboundary Movements of Hazardous Wastes. https://wedocs.unep. org/bitstream/handle/20.500.11822/8385/-Basel%20Convention%20on%20the%20Control%20of%20Transboundary%20 Movements%20of%20Hazardous%20Wastes%20-20113644.pdf?sequence=2&%3BisAllowed=.

[43] Green Cooling Initiative. 2024. *Green Cooling Technologies.* https://www.green-cooling-initiative.org/green-cooling/ technology.

bridge the gap between current costs of green cooling technologies and current practices.[44] The project will establish the necessary capacity and regulatory conditions for the adoption of green cooling, resulting in the crediting of achieved mitigation and transfer of credits through Article 6.2 of the Paris Agreement.

Successful experiences under Article 6 might also boost buyer awareness for this project type, making it possible to leverage the voluntary carbon market (VCM) as an effective tool for promoting green cooling technologies.

In future evolutions of carbon market methodologies, such as through the development or revision of existing methodologies, carbon markets may also leverage passive cooling and architectural solutions to create green cooling solutions. This involves architectural design elements and solutions inspired by passive cooling solutions and traditional knowledge that has been applied historically in various regions of the world.[45]

(v) Leak detection and prevention

Although a refrigerant leak detection methodology from Verra has been in place since 2012, only one project so far has been registered applying it. However, the methodology has been recently revised in 2024, to expand the types of technologies that are eligible. This may lead to an increased uptake in the use of the methodology. Given the limited uptake of the methodology to date, it is complex to identify use cases or further information about applicability and prices. Nevertheless, the project type, namely leak detection and prevention, remains a key part of LRM adoption, and further work in the area can lead to effective LRM solutions. Furthermore, given the potential effectiveness of such projects in achieving greater LRM implementation, policymakers and prospective project developers may look toward evaluating the potential of the project type within their national context on a case-by-case basis, including under Article 6 of the Paris Agreement.

Availability of Destruction or Reclamation Facilities

Depending on the national context, facilities destroying or reclaiming fluorocarbons may or may not exist nationally; therefore they may or may not be available for implementing LRM actions and carbon markets. Article 5 countries may find different national circumstances depending on their historical engagement with fluorocarbon destruction or reclamation activities.

Carbon markets, however, may be leveraged themselves to promote the construction or retrofitting of such a facility, thereby supporting a potential strategic objective of participation and enabling LRM nationally. To date, existing private sector project developers and other market participants have indeed successfully financed and constructed specialized infrastructure for the purpose of developing and operating carbon market projects destroying or reclaiming refrigerant gases. In cases where such facilities are not available, market participants have also resorted to the export of gases as part of carbon market projects.

COPA identifies several reclamation technologies employed by reclamation facilities for fluorocarbon gases, including distillation, adsorption, and subcooling. In addition, the report identifies several specific destruction technologies, including cement kilns, municipal solid waste incineration, rotary kiln incineration, and argon plasma arc. Of these, cement kilns and municipal solid waste incineration facilities may be most

---

[44]   Carbon Mechanisms Group. 2024. *Cooling Programme Southern Africa*. https://www.carbon-mechanisms.de/en/news-details/cooling-programm-southern-africa.

[45]   M. A. Kamal. 2012. An Overview of Passive Cooling Techniques in Buildings: Design Concepts and Architectural Interventions. *Acta Technica Napocensis: Civil Engineering & Architecture*. 55(1). https://www.solaripedia.com/files/1388.pdf.

accessible and feasible for implementation in Article 5 countries, with lower costs, potential to retrofit existing infrastructure, and potentially high efficiency. TEAP in its May 2024 report further details the costs, barriers, and technical features of both reclamation and destruction technologies and facilities.

Stakeholders, including private sector project developers or public entities, will be incentivized when they engage in the retrofitting or construction of such a reclamation or destruction facility if several criteria are met, which make such an undertaking financially viable. The following criteria are often critical for determining the viability of constructing or retrofitting facilities:

- Carbon credit price ($/tCO$_2$e) obtained for destruction or reclamation activities
- Carbon credit demand and prospective credit buyers
- Volume of gases available for destruction or reclamation, including from potential collection efforts, at present and over longer periods of time, informing the economic viability of such a facility over time
- Price and demand for reclaimed refrigerants (in the case of reclamation)
- Price of virgin refrigerants (in the case of reclamation)
- Availability of existing infrastructure for retrofitting (e.g., cement kilns, in the case of destruction)
- Possibility to obtain corresponding adjustments under Article 6 of the Paris Agreement, and availability of bilateral agreements, Article 6 frameworks and regulations, and other Article 6 policies and decisions

In many cases, stakeholders, including project developers and financial entities, are already evaluating these criteria to identify potential projects and determine investment opportunities.

Countries and policymakers may wish to understand the criteria included above in order to be better able to promote positive outcomes or design policies and decisions that may catalyze the construction of essential fluorocarbon reclamation or destruction infrastructure. While safeguarding the importance of additionality and environmental integrity, policymakers may take concrete steps toward the success of LRM approaches by making strategic decisions that can influence the criteria above. This may involve, for example, communicating willingness to offer corresponding adjustments to high-integrity projects that engage in the construction or retrofitting of facilities for fluorocarbon management.

Furthermore, different international development partners, bilateral partners, or crediting mechanisms may further offer upfront financing, which may be leveraged for the construction of needed infrastructure. An example of this is Japan's Joint Crediting Mechanism wherein financing may be provided for model projects for a portion of investment costs, such as those needed to cover the construction of facilities.[46] ADB also operates the Japan Fund for the Joint Crediting Mechanism which includes grants for the construction of infrastructure in energy-related projects and for additional and advanced technology components.[47] Broadly, stakeholders including policymakers and project developers may look toward similar opportunities to leverage the financing for the construction of fluorocarbon management facilities.

---

[46] Joint Crediting Mechanism. Preliminary Selection Result for Financing Programme for JCM Model Projects in FY2023 (1st Selection). https://gec.jp/jcm/kobo/mp230817/.

[47] ADB. Japan Fund for the Joint Crediting Mechanism. https://www.adb.org/what-we-do/funds/japan-fund-for-joint-crediting-mechanism.

2.3.2 Identifying and Evaluating Projects as a Country Government

Once the national context is clearly identified and strategic objectives are set, the next step is to develop a project pipeline to identify potential fluorocarbon life cycle management activities that could be financed by carbon markets. Countries may choose from a variety of approaches to create a pipeline, ideally involving a close collaboration between the ozone and climate units. Some of the approaches that countries may use are as follows:

- Launch a call for proposals aligned with the strategic objectives. Governments may choose to launch a public call for proposals requesting stakeholders to submit project ideas and proposals for fluorocarbon projects.

- Engage with stakeholders and entities responsible for various sources of emissions via workshops, roundtables, and conferences. Governments may leverage known networks or other government entities that may have close links to potential sources of project ideas.

- Identify project developers that have national or international experience in the project type. International stakeholders and private sector project developers may be well connected to project ideas and knowledgeable about national circumstances.

- Identify and prepare a project through dialogue with development partners such as ADB. While countries may independently identify projects, development partners are often well suited to identifying projects, supporting with international experience and technical experts, and supporting project design.

- Engage with prospective buyers of units. Actors on the demand side can help inform national strategies by sharing perspectives on future credit needs, standards, and expected co-benefits, among others.

When identifying potential projects within a country, it will be also relevant to investigate what other projects in the pipeline are (e.g., projects under other financing mechanisms such as international cooperation). One such example is the Multilateral Fund for the Implementation of the Montreal Protocol (MLF) of which the main objective is to provide financial and technical assistance to Article 5 countries to help them comply with their obligations under the Protocol. Such activities, which aim to phase out ODS or phase down HFCs, may include closure of ODS production plants, industrial conversion, technical assistance, information dissemination, training, and capacity building. How the MLF will finance ODS and HFC disposal projects, such as destruction facilities, is currently under discussion.[48] To qualify for this type of funding, parties would need to submit a national ODS/HFC bank inventory to accurately estimate mitigation potential.[49]

Projects Funded by the Multilateral Fund

The MLF provides funding in the form of grants. In the case of technology conversions, it only funds the additional or "incremental" costs incurred. The eligibility of MLF-funded activities for carbon credits depends primarily on additionality, which needs to be assessed on a case-by-case basis. In general, if the MLF funding covers incremental costs making the project activity economically viable, it is most likely that there is no additionality justification, and therefore there is no possibility to obtain carbon credits.

---

[48] United Nations Environment Programme. 2024. Decision XXVIII/2: Decision Related to the Amendment for Phasing Down Hydrofluorocarbons. https://ozone.unep.org/treaties/montreal-protocol/meetings/twenty-eighth-meeting-parties/decisions/decision-xxviii2-decision-related-amendment-phasing-down-hydrofluorocarbons.

[49] United Nations Environment Programme. 2024. *MOP 35 Decision 12: Report of the Twenty-Fifth Meeting of the Parties to the Montreal Protocol.* https://ozone.unep.org/system/files/documents/MOP-35-12-Add-1E.pdf.

Moreover, if the MLF funding supports an activity intended to achieve compliance, it is unlikely to be additional unless it can be demonstrated that the activity leads to accelerated compliance with the Montreal Protocol. On the other hand, if the MLF funding is not enough to overcome all financial and other barriers and goes above and beyond compliance, there may be an opportunity to receive cofinancing via carbon credits. In summary, as with all cases of cofinancing, a thorough evaluation of each project's additionality will be necessary to determine whether it is eligible for carbon finance.

2.3.3 Evaluating Project Eligibility and Viability

Once potential projects are identified, the next step is a screening process to check whether the projects are potentially eligible for carbon credits, followed by an estimate of the project's potential to generate emission reductions. An early estimate (often very rough at the beginning) gives an indication whether the project is financially viable through carbon revenues. Often a breakeven cost is estimated and compared to the market sales price of emission reductions. Both figures often involve a lot of assumptions and face a lack of available data, making these cost–benefit estimates in the early days challenging.

Table 3 presents a list of some criteria and categories of information that may be screened or evaluated when assessing project eligibility and viability. This list is not exhaustive and may need to be adapted based on specific country conditions or project types, particularly given the dynamic and evolving nature of the market for new green cooling technologies.

Table 3: Criteria and Categories of Information to Screen or Evaluate for Project Eligibility and Viability

Category	Description
Source of gas	• Where is the gas coming from (which equipment)? • What is the level of ease in collection? • Is a collection system in place?
Type and amount of gas	• Which gases are available? • Are they ODS or HFCs? • What is their GWP? • What amounts are available, and within what time frame?
Destruction/reclamation facilities	• Are there facilities existing in the country? • What are their technical specifications? • Do they comply with TEAP and/or other national requirements? • What are the feed/reclamation rates?
Baseline scenario analysis	• What is the expected scenario in the country without carbon credits? • What is the existing national legal and regulatory framework? • What are the national policies on LRM? • Is the country an Article 5 country and what is the country's phasedown schedule? • What are the economic/technological circumstances in the country around LRM?

continued on next page

Table 3 *continued*

Category	Description
Additionality assessment	• What is the common practice in the country? • Is the project financially attractive without carbon revenues? • What regulations, if any, already mandate the activity to be covered by the project? • Are there other barriers to the implementation of the project beyond financial ones, such as lack of skilled technicians, lack of facilities, and difficulties in acquiring sufficient amounts of gases?
Monitoring	• Do accredited laboratories that can carry out the required analyses and tests exist in the country? • Is an electronic database or registry available for tracking and monitoring fluorocarbons? • Is information on the chain of custody available?

GWP = global warming potential, HFC = hydrofluorocarbon, LRM = life cycle refrigerant management, ODS = ozone depleting substance, TEAP = Technology and Economic Assessment Panel.

Source: Asian Development Bank.

2.3.4 Applicability of Methodologies

When designing high-integrity carbon credit projects, it is essential to use robust, conservative accounting methodologies that are clearly and transparently documented. The fragmentation of the international carbon markets and the lack of a central regulator has resulted in a variety of different entities developing methodologies, such as independent standards, the UNFCCC, protocols developed for regional compliance schemes such as the California Cap-and-Trade Program, or proactive government initiatives such as Japan's Joint Crediting Mechanism. The cooperative approaches under Article 6.2 are more flexible than the VCM or previously CDM because, in principle, they allow the use of any methodology agreed upon bilaterally between the host country and the buyer country.

On the other hand, the Paris Agreement Crediting Mechanism only allows the crediting of projects that apply methodologies approved by the UNFCCC for use under Article 6.4. Similarly, the carbon credits that airlines can purchase to meet their CORSIA emission reduction requirements must be issued by a CORSIA-approved standard, and need to fulfill specific eligibility criteria such as project type, vintage year, and having received corresponding adjustments.

Furthermore, some buyer countries have also placed requirements on the types of credits they are willing to authorize, including specifying which standards they will accept. For instance, Singapore has a carbon tax mechanism that is applied to certain industrial facilities and allows these to use authorized ITMOs from specific standards and project types to offset up to 5% of their taxable emissions.[50]

For projects developed under the VCM, the eligibility of ODS and HFC projects is highly dependent on the availability of methodologies with applicability criteria that fit the type of gases and type of intervention or location. The very specific requirements in these methodologies are sometimes difficult for projects to comply with, such as being eligible only for certain geographic locations or specifications on the destruction technology (e.g., TEAP). Generally, project developers will closely review methodologies to establish whether project ideas can be successfully implemented based on the requirements set out in them.

---

[50] National Environment Agency. 2024. *Carbon Tax.* https://www.nea.gov.sg/our-services/climate-change-energy-efficiency/climate-change/carbon-tax.

The following list summarizes carbon project methodologies for addressing refrigerant emissions under various standards, including multiple project types:

ODS Destruction

- **California Air Resources Board** has a Compliance Offset Protocol for ODS destruction projects in which the ODS must originate in the US.[51]

- **VERRA** has a methodology that can be applied to projects recovering and destroying ODS from products internationally.[52]

- **Climate Action Reserve** has three protocols for ODS destruction for ODS sourced from the US, Article 5 countries, or Mexico. The first two require that the ODS is destroyed in the US, whereas the latter—the Mexico Halocarbon Protocol—is for ODS/HFC sourcing and destruction in Mexico.[53]

- **American Carbon Registry (ACR)** has methodologies for ODS destruction both from the US and international sources.[54]

HFC Destruction

- **Global Heat Reduction Registry**, which was launched in 2024, announced a new methodology for HFC destruction on 23 September 2024.[55]

- **Carbon Containment Lab** has published a draft methodology for the destruction of HFCs in Article 5 countries.[56]

HFC Reclamation

- **ACR** has a methodology aimed at reducing greenhouse gas (GHG) emissions through the use of certified reclaimed HFC refrigerants, propellants, and fire suppressants to displace the production and use of virgin HFC gases for projects located in the US, Canada, or Mexico.[57]

Green Cooling

Green cooling covers the transition to the use of natural refrigerants in combination with highly energy-efficient appliances. Although this is the ultimate goal of decarbonization in the refrigeration and air-conditioning sector, the transition is challenging and the carbon market has been slow in developing new methodologies in this sector, with few approved methodologies to date. Initiatives led by GIZ in Ghana and South Africa to develop green cooling projects under the Article 6.2 mechanism are at early stages but may serve as models for future projects. Further work on new methodologies will nevertheless be necessary to enable access to climate finance for transforming the sector.

---

[51] California Air Resources Board. 2014. *Compliance Offset Protocols: Ozone Depleting Substances Projects.* https://ww2.arb.ca.gov/our-work/programs/compliance-offset-program/compliance-offset-protocols/ozone-depleting-substances-projects.

[52] Verra. 2017. *VM0016: Recovery and Destruction of Ozone-Depleting Substances (ODS) from Products, v1.1.* https://verra.org/methodologies/vm0016-recovery-and-destruction-of-ozone depleting-substances-ods-from-products-v1-1/.

[53] Climate Action Reserve. 2012. *Ozone Depleting Substances Protocol.* https://www.climateactionreserve.org/how/protocols/industrial/ozone-depleting-substances/.

[54] American Carbon Registry. 2024. *Approved Methodologies.* https://acrcarbon.org/methodologies/approved-methodologies/.

[55] Global Heat Reduction Registry. 2024. *Global Heat Reduction Registry Announces First Three Methodologies/Projects.* https://www.heatreduction.com/news/press-release/global-heat-reduction-registry-announces-first-three-methodologies-projects.

[56] Carbon Containment Lab. 2024. *HFC Methodology.* https://carboncontainmentlab.org/publications/hfc-methodology.

[57] American Carbon Registry. 2022. *Certified Reclaimed HFC Refrigerants, Propellants, and Fire Suppressants Methodology.* https://acrcarbon.org/methodology/certified-reclaimed-hfc-refrigerants-propellants-and-fire-suppressants/.

■ **ACR** has a methodology for projects that avoid the emissions of CFC, HCFC, or HFC gases through the deployment of advanced refrigeration systems in commercial refrigeration for projects located in the US, Canada, or Mexico.[58]

■ **UNFCCC Clean Development Mechanism** has a number of methodologies that may be applicable to these project types, however, some of them only address energy savings.[59] These include the following:

 □ AMS III.X Energy Efficiency and HFC-134a Recovery in Residential Refrigerators

 □ AMS III.AB. Avoidance of HFC emissions in Standalone Commercial Refrigeration Cabinets

 □ AMS-II.C. Demand-side energy efficiency activities for specific technologies

 □ AMS-II.O. Dissemination of energy efficient household appliances

 □ AM0060 Power saving through replacement by energy-efficient chillers

 □ AM0070 Manufacturing of energy-efficient domestic refrigerators

 □ AM0071 Manufacturing and servicing of domestic and/or small commercial refrigeration appliances using a low GWP refrigerant

 □ AM0120 Energy-efficient refrigerators and air conditioners.

Leak Detection and Prevention

■ **Verra** has a methodology for the installation of infrared automatic leak detection systems in commercial refrigeration systems using HFCs.[60] This methodology measures the reduction in emissions achieved by reducing HFC leaks from commercial refrigeration systems in the US.

2.3.5 Integrity and High-Quality Carbon Credits

Developing high-integrity methodologies and projects is crucial for carbon market mechanisms to deliver real, verifiable climate impacts. However, given the voluntary and self-regulating nature of the VCM, the lack of consistency and the fragmentation in the market has undermined the credibility of the system. Nevertheless, there are ongoing initiatives to ensure a consistent level of quality, such as the Integrity Council for the Voluntary Carbon Market and the Voluntary Carbon Markets Integrity Initiative.

Carbon credits generated from fluorocarbon mitigation are often highly additional since the underlying activities, without carbon finance, are typically unprofitable (especially end-of-life projects) and are permanent (e.g., a destructed gas cannot be reversed). Furthermore, LRM projects also deliver various co-benefits beyond climate impacts, such as sustainable development through the creation of skilled jobs and improved waste management systems. However, there are potential risks associated with these project types. This section summarizes some key elements that need careful consideration to ensure high integrity in LRM carbon credits.

---

[58] American Carbon Registry. 2021. *Advanced Refrigeration Systems Methodology.* https://acrcarbon.org/methodology/advanced-refrigeration-systems/.

[59] UNFCCC. 2024. *Clean Development Mechanism Methodologies.* https://cdm.unfccc.int/methodologies/index.html.

[60] Verra. 2024. *VM0001 Refrigerant Leak Detection, v1.2.* https://verra.org/methodologies/vm0001-refrigerant-leak-detection-v1-2/.

Prevention of Perverse Incentives

One of the key lessons learned from the Kyoto Protocol's Clean Development Mechanism (CDM) is the need for methodologies to include appropriate safeguards to prevent any perverse incentives. It was reported that in certain HFC-23 destruction CDM projects, the amount of HFC-23 (a by-product of HCFC-22 production) was intentionally increased to increase the number of credits, which would mean that the CDM provided a perverse incentive to generate more HFC-23.[61] Therefore, methodologies must have appropriate safeguards to prevent such perverse incentives, such as requiring the use of a capping factor that is based on historic production figures combined with an ambitious Kigali phasedown schedule for the baseline emissions. Another concern is that a country's fear of jeopardizing the additionality of potential projects should not delay the introduction of national regulations. Activities are deemed additional only if there is no enforced national regulation governing the specific activity type, such as Extended Producer Responsibility (EPR) schemes or other enforced waste management policies.

Ambitious Baseline Setting

Another potential risk is the over-crediting of emission reductions due to inaccurate or weak baseline setting. Emission reductions are calculated as the difference between the emissions in a hypothetical baseline scenario versus the actual emissions from the project activity. Figure 13 illustrates a typical baseline and project scenario.

However, if the hypothetical baseline predicts much higher emissions than what actually would have happened in the absence of the project, this could result in an overestimation of carbon credits. Refrigerant projects need to consider carefully the phaseout or phasedown schedules under the Montreal Protocol when setting the baseline. For example, a project design may include a baseline calculation that transparently includes how the baseline takes into account the expected reductions in emissions arising from the implementation of the Montreal Protocol.

Another potential risk in identifying baselines is that fluorocarbons are exported from one country that have a legislation requiring destruction or reclamation, to another that does not have such legislation, thereby overestimating the baseline emissions. Another risk is that the destruction of HFCs, if conducted in compliance with TEAP destruction methods, entitles the country to add these amounts to their consumption allowance under the Montreal Protocol. These gases may then be used in equipment outside of the project scope, thus potentially causing emissions that may not be accounted for under the project. Both of these risks should be considered when identifying the baseline and may be a part of methodology requirements or project design and validation.

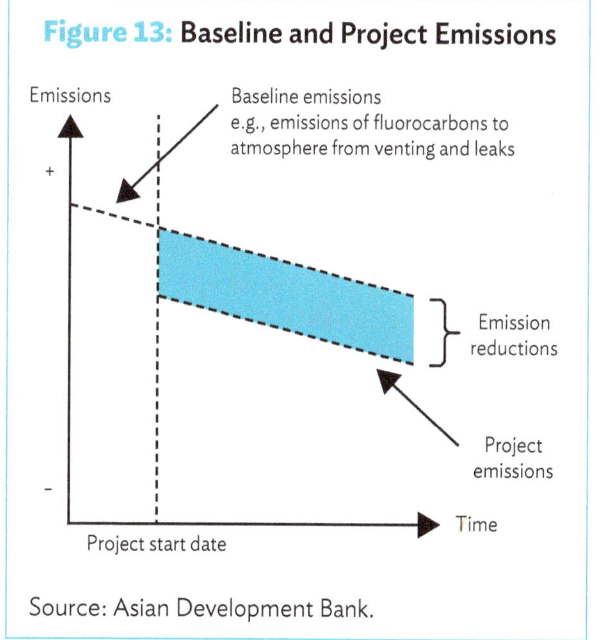

Figure 13: Baseline and Project Emissions

Source: Asian Development Bank.

---

[61]   Deutsche Gesellschaft für Internationale Zusammenarbeit (GIZ). 2015. *Management of Ozone-Depleting Substances (ODS) Banks: A Study*. https://www.giz.de/en/downloads/giz2015-en-study-ods-banks-management.pdf.

National circumstances and context are highly relevant to identifying baseline emissions and may also be included in explorations of the baseline. Baseline setting may vary considerably between project types, crediting methodologies, and other factors. Overall, baseline determination should consider the project type, methodology, national context, and other factors. Appropriate risk-safeguarding measures should also be in place, such as determining a list of applicable methodologies, or criteria for the authorization of projects that include conservative baseline setting.

In a nutshell, ambitious baseline setting is required to ensure conservative calculations of emission reductions, including a thorough investigation of legal frameworks (e.g., Montreal Protocol, Paris Agreement, and relevant national legislation). The 2023 COPA paper on ODS/HFC policy recommendations gives further insight into these integrity issues and provides recommendations for developing high-integrity carbon projects in the refrigeration and air-conditioning sector.[62]

While the determination of baselines is a central element to crediting projects under the carbon markets, the responsibility for determining a baseline and how it may be calculated is oftentimes split. In most cases, the rules for determining a baseline and the data that must be taken into account is set by a carbon standard within a crediting methodology. The responsibility for carrying out the calculations and modeling required to establish the baseline itself often lies with the project design and project developer teams, and may vary depending on national context, project type, gas, scope, and region, among others. Overall, countries may look to understand where they strategically choose to provide inputs and data, regulate, or be directly involved in the process of establishing a baseline for a particular project or project type. Depending on in-house capacity needs or other factors, the task may be best delegated to other relevant stakeholders.

Regulatory Additionality and Delayed Implementation Risks

Maintaining environmental integrity is essential in all carbon finance aspects and in carbon project development. In cases where national regulations mandating the destruction of fluorocarbons or various other approaches to the life cycle management of fluorocarbon projects may be in place, an important facet to maintaining environmental integrity is ensuring that projects requesting carbon finance for their operations would have not occurred otherwise, due to the implementation of such national regulations. In multiple countries today, there are examples of national or other regulations that mandate that fluorocarbon gases are addressed in some form, such as regulations mandating the destruction of gases.

Historically, this risk has been addressed as part of "regulatory additionality" or "regulatory surplus" provisions in standards or methodologies, that is, that any given project is additional to (not mandated by) national regulations, requirements, or laws. In principle, the concept of regulatory additionality requires that a project is not mandated by law in order to be eligible for carbon finance, mitigating the risk that a project that requests carbon credits would have occurred anyway in the absence of carbon finance, given existing national regulations mandating it. In all carbon projects, this is an important provision to ensure environmental integrity and that carbon finance only goes to projects that require this finance in order to operate.

---

[62] Climate and Ozone Protection Alliance (COPA). 2023. *ODS/HFC Reclamation and Destruction Technologies: A Review for Article 5 Countries.* https://www.copalliance.org/imglib/publications/COPA%20Publication%20Policy%20Study%202023.pdf.

However, oftentimes regulations are not followed due to a lack of enforcement, a lack of monitoring, a lack of incentives, or a variety of other barriers to carrying out the actions mandated by regulations, on top of the high complexity of various sectors. Within the fluorocarbon sector, this is particularly prevalent, with a wide range of highly developed countries still only achieving modest levels of compliance with regulations and of reclamation or destruction, at various points in the supply chain of fluorocarbons. Furthermore, the enforcement of such regulations within the fluorocarbon sector often becomes even more difficult because of the complexity of identifying the points at which, for example, gases have leaked during the equipment's lifetime. This is the case for multiple actions comprising LRM, where even despite regulations preventing venting of gases and mandating their collection, gases are often vented at servicing or during the replacement of equipment.

To address this, a variety of carbon project methodologies already include provisions that enable the project to be eligible if sufficient proof is provided that a regulation is not being actively followed in the sector broadly. These provisions may include thresholds which may apply nationally or regionally, such as the rate of fluorocarbon destruction or reclamation in a particular country or region. If a project is able to provide sufficient evidence, in the form of data or other publications, demonstrating that the particular national context or situation is below such a threshold, and that national regulations are not enforced or followed, then the project may be deemed additional. Setting such a threshold for project eligibility, based on compliance rate with a policy or reclamation or destruction rates nationally, is very context-dependent and is therefore highly recommended to be determined nationally.

Project proponents may also need to draft an explanatory argument for the project meeting regulatory additionality criteria based on national context, supported by adequate evidence. The data and argumentation provided will also be validated and verified by third-party validation and verification bodies (VVBs). Ensuring that these provisions are working effectively and enabling projects that are truly additional and would not have occurred in the absence of carbon finance is essential to maintaining high-integrity projects. Furthermore, countries may also choose to include items as part of the eligibility provisions, whether at an authorization stage or other, by adding a criterion on regulatory surplus.

Another important risk to be mitigated, within the context of regulatory additionality, is the concept that carbon finance may provide an incentive to delay the implementation of legislation, policies, or regulations that would have otherwise been implemented. Such a risk may occur when carbon finance is deemed exceedingly attractive, causing a country government to delay, cancel, or not pursue the operationalization of a policy that would otherwise have been implemented in some form. One way country governments may address this risk is by transparently communicating climate policies and intended actions, highlighting the barriers that they may face in operationalizing such climate actions without relying on carbon finance. This may be done, among other ways, by communicating intended actions as part of their NDC commitments. Addressing the risk of delaying policy or regulation implementation promotes environmental integrity and helps ensure that projects have a strong case for carbon finance.

Permanence

Permanence is an important factor in environmental integrity and its absence hinders the ability of a project to make lasting climate impacts. On the one hand, fluorocarbon projects that destroy fluorocarbons have a strong permanence argument—gases are permanently destroyed and therefore are not at a risk of being emitted into the atmosphere, that is, for the emission reductions that were credited to be effectively reversed. On the other hand, the gases reclaimed in reclamation projects, which are put back in use, may still vent to the atmosphere at later stages, thereby reversing some of the emission reductions calculated by the project.

While permanence is a risk to consider in reclamation projects, projects crediting emission reductions arising from the reclamation of fluorocarbons may still have effective climate impacts and may support managing LRM in countries. In fact, a range of carbon accounting methodologies exists for projects reclaiming fluorocarbons. Overall, this permanence risk may not directly determine the eligibility of reclaiming projects, but understanding the risk and taking it into account in the design of project eligibility criteria may be an important step to ensure that environmental integrity is balanced with the leveraging of carbon finance opportunities.

The implementation of LRM approaches and specific project design features can play a strong role in ensuring that the emission reductions of a project are permanent. For example, EPRs and similar features may be included in the project design so that reclaimed gases are monitored throughout their life cycle, and the risk of venting or leaking once reclaimed gases are put into equipment is minimized. A project design may also include implementing measures for the traceability of fluorocarbons in equipment, such as through the collection of information of installed equipment and software in order to monitor the life cycle of a fluorocarbon.

2.3.6 Finding a Buyer

In the context of Article 6, ITMOs may be bought and traded by buyers for a range of purposes. One of the primary uses of ITMOs is their eligibility to be counted toward NDC targets. As such, multiple countries to date have expressed their interest in the purchase of ITMOs toward the achievement of national targets.

Governments identifying buyers for fluorocarbon projects must understand buyer requirements. First and foremost, buyers are looking at environmental integrity and the transparency of the positive climate impact and co-benefits of a project. Before finding a buyer, governments and private sector developers must consider buyer needs regarding environmental integrity, including among others the following criteria:

- Strong additionality
- Conservative and transparently derived baselines, such as the integration of Kigali phasedown schedules
- Adequate project monitoring plans and publicly available emission reduction calculations
- Safeguards are in place to prevent perverse incentives
- Adequate sustainability and societal safeguards
- Clear information about co-benefits, such as training technicians, job creation, waste management strengthening, and improving data collection and national inventories
- Robust monitoring, reporting, and verification (MRV) systems

Understanding market demand for ITMOs is key to determining the potential for projects at early stages to establish a financing stream through this mechanism. Only Article 6.2 is currently operational. Therefore, the potential of developing a project under Article 6.2 hinges on buyer country governments, private sector buyers, and other international organizations purchasing credits, as well as host countries willing to authorize credits. One example of other parties interested in purchasing ITMOs, beyond country governments, may be airlines who are interested in buying ITMOs to fulfill their obligations under CORSIA, which is expected to be a significant source of demand.[63]

---

[63]   E. Gupte. 2024. *Demand for CORSIA Carbon Credits Unlikely to Match Supply Until 2030: Abatable. S&P Global Commodity Insights*. https://www.spglobal.com/commodityinsights/en/market-insights/latest-news/energy-transition/051424-demand-for-corsia-carbon-credits-unlikely-to-match-supply-until-2030-abatable#:~:text=In%20a%20report%2C%20Abatable%20modeled,to%20meet%20demand%20until%202029.

Countries transacting ITMOs from government to government sign bilateral agreements to formalize their cooperation and establish ground rules for the development of projects, project types, and transactions. The number of bilateral agreements between parties, pledges to purchase ITMOs, and host country governments willing to authorize credits is slowly rising.[64] Also, the number of governments announcing intentions to purchase credits and to develop national frameworks and regulations enabling Article 6 projects is also increasing.

To date, the following countries have signed bilateral agreements as likely buyers and are currently purchasing ITMOs or are likely to do so in the short or medium term:

- Japan (through the Joint Crediting Mechanism)
- Kuwait
- Norway
- Singapore
- Republic of Korea
- Sweden
- Switzerland
- United Arab Emirates

Overall, there is still modest demand in the market for carbon credits authorized under Article 6. However, the market is still nascent, and in the future, it is likely that commitments by country governments, coupled with demand rising from the private sector, can positively impact demand.

In the context of the VCM, it remains challenging to find buyers beyond North America that are willing to pay prices high enough to make projects viable. This is partly due to the lack of information and awareness among buyers of the high global warming impacts and mitigation potential in these project types, along with a misconception that these do not come with "sufficient" sustainable development co-benefits. Buyers tend to pay higher prices for projects that demonstrate co-benefits and tell charismatic stories, which may sometimes be more difficult to portray in the refrigeration and air-conditioning sector. Furthermore, the experiences in the CDM with HFC-23 destruction projects have also led buyers to misunderstand these project types and that appropriate safeguards can easily mitigate such risks.

Thus, project developers may need to spend much time educating buyers and explaining the high impact surrounding these projects, showcasing their sustainable development co-benefits such as training a skilled workforce and establishing robust MRV systems and explaining that lessons learned from the CDM have resulted in very stringent safeguards and robust accounting mechanisms.

2.3.7 Policy Crediting Approaches in Life Cycle Refrigerant Management

Policy crediting approaches are an alternative to project-by-project crediting which is the norm in carbon finance since the Kyoto Protocol. They function by quantifying and crediting the mitigation impact of the implementation of a particular policy and have been discussed by mitigation and policy experts intensively, but to date only a few examples exist. In part, this is due to the complexity of implementing such a policy approach and quantifying it transparently and accurately. However, beyond the challenges

---

[64]   United Nations Environment Programme. 2024. *Article 6 Pipeline.* https://unepccc.org/article-6-pipeline/.

that do exist, policy crediting approaches have greater potential to level up the mitigation of emissions by providing finance to vast swathes of emissions at a different scale compared to individually crediting project-by-project approaches. Literature on policy crediting approaches has identified a range of different implementable policies (Figure 14).

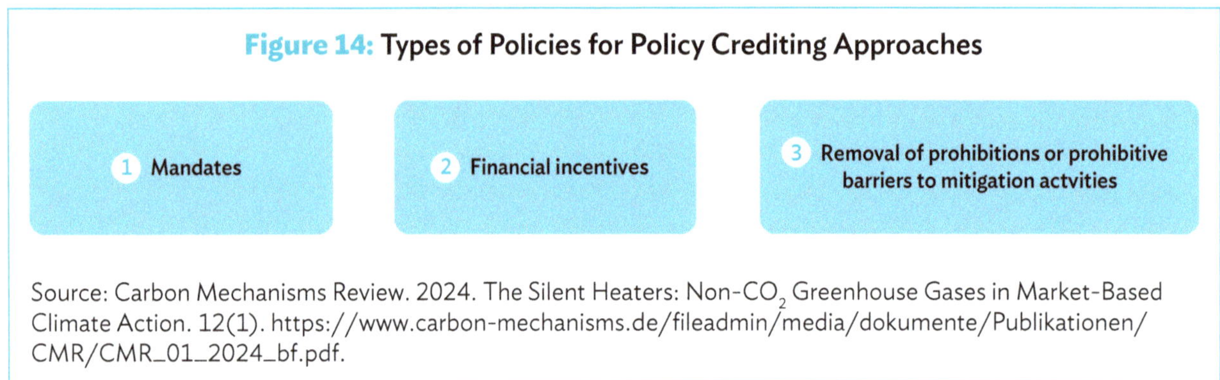

Figure 14: Types of Policies for Policy Crediting Approaches

Source: Carbon Mechanisms Review. 2024. The Silent Heaters: Non-CO$_2$ Greenhouse Gases in Market-Based Climate Action. 12(1). https://www.carbon-mechanisms.de/fileadmin/media/dokumente/Publikationen/CMR/CMR_01_2024_bf.pdf.

This publication does not explore policy crediting approaches in detail due to their complexity, potential risks, and the careful planning required for successful implementation. However, given their significant potential for emission reduction and for transforming the sector on a large scale, this area should be a primary focus for in-depth research and investigation in future studies.

2.4 Accounting and Reporting for Fluorocarbon Projects Within the Montreal Protocol and Paris Agreement

Accounting and reporting requirements are of particular importance within the context of fluorocarbon management. Countries that employ Article 6 carbon markets to address LRM need to be aware of them.

This section describes transparency, accounting, and reporting requirements under Article 6 within the context of fluorocarbon projects.

2.4.1 Transparency, Accounting, and Reporting Requirements Under Article 6 and Article 13 of the Paris Agreement

The Paris Agreement, under Article 13 Enhanced Transparency Framework, mandates the submission of National Inventory Documents (NIDs) covering emissions and removal of the following greenhouse gases (GHGs): carbon dioxide, methane, nitrous oxide, hydrofluorocarbons, perfluorocarbons, sulfur hexafluoride, and nitrogen trifluoride.[65] Transparency refers to the reporting of information by a signatory to the Paris Agreement to track NDC progress and provide information on the support mobilized, needed, and received as part of its commitments and targets. Its primary vehicle is the Biennial Transparency Report (BTR), under which the national inventories are reported.

---

[65]   UNFCCC, CMA. 2016. Decision 18/CMA.1, Annex, Paras. 1, 2, 146, and 189. Modalities, Procedures and Guidelines for the Transparency Framework for Action and Support Referred to in Article 13 of the Paris Agreement. https://unfccc.int/resource/tet/0/00mpg.pdf.

NDC accounting, on the other hand, refers to the processes, rules, and principles applied by parties to the Paris Agreement in tracking progress toward their NDCs, in line with the guidance set out. Within their BTRs, countries must include information about their participation in cooperative approaches under Article 6, and therefore must include information about any participation in activities generating ITMOs, including those related to fluorocarbons.

When implementing Article 6 projects in fluorocarbon management, countries must submit several key documents as illustrated in Figure 15.

Figure 15: Information Submitted as Part of the Initial Report, Annual Information, and Regular Information

Initial Report	Annual Information (AEF)	Regular Information (BTR)
The Party must demonstrate that it • Is a Party to the Paris Agreement; • Has prepared, communicated, and is maintaining an NDC; • Has arrangements in place for authorizing the use of ITMOs toward the achievement of NDCs; • Has arrangements in place for tracking ITMOs; • Has provided the most recent national inventory report required in accordance with decision 18/CMA.1; and • Can ensure its participation contributes to the implementation of its NDCs and long-term low-emission development strategy, if it has submitted one, and the long-term goals of the Paris Agreement.	**Annual information through the AEF on** • Authorization of ITMOs for use toward achievement of NDCs; • Authorization of ITMOs for use toward Other International Mitigation Purposes; and • First transfer, transfer, acquisition, holdings, cancellation, voluntary cancellation, voluntary cancellation of mitigation outcomes or ITMOs toward overall mitigation in global emissions, and use toward NDCs.	In addition to restating and **updating any information submitted as part of the initial report**, Parties must also provide information on the **authorization, corresponding adjustments, and use of ITMOs**. Furthermore, they must include **information on each cooperative approach**.

AEF = Agreed Electronic Format, BTR = Biennial Transparency Report, CMA = Conference of the Parties serving as the Meeting of the Parties to the Paris Agreement, ITMO = internationally transferred mitigation outcome, NDC = nationally determined contribution.

Source: UNFCCC, CMA. 2021. Decision 2/CMA.3, Annex, Section IV – Reporting, Guidance on Cooperative Approaches Referred to in Article 6, Paragraph 2, of the Paris Agreement. https://unfccc.int/sites/default/files/resource/cma2021_10_add1_adv.pdf.

While this document will not go into detail about how countries may develop and submit the above information, the key links between fluorocarbon projects and the initial report, annual information, and regular information to be submitted are highlighted below.

Accounting and Reporting of Hydrofluorocarbons

While developed countries are required to report HFC emissions in their NIDs, developing countries are encouraged, but are not obligated, to do so.[66] Notably, ODS (CFCs and HCFCs) are not included in the list of gases that must be reported.

---

[66] UNFCCC, CMA. 2016. Decision 18/CMA.1. Modalities, Procedures and Guidelines for the Transparency Framework for Action and Support Referred to in Article 13 of the Paris Agreement. https://unfccc.int/resource/tet/0/00mpg.pdf.

Given that HFCs are covered under the Paris Agreement and under the 2006 IPCC Guidelines for National Greenhouse Gas Inventories,[67] they are accounted and reported as GHG units measured in tonnes of carbon dioxide equivalent (tCO_2e). This section describes the accounting and reporting from the perspective of HFC projects.

Annual Information

Countries developing HFC management projects must account for all emission reductions arising from the project as part of the methodology employed. Based on Decision 2/CMA.3, annex, paras. 20(a) and 20(b), and the draft version of the Agreed Electronic Format,[68] countries must submit detailed project-based information[69] in three tables:

- Table 1—Heading: Information on the submitting Party and the reported year

- Table 2—Actions: Detailed information on all actions (transfers, authorizations, uses, and cancellations) that occurred within the year from 1 January to 31 December of the reported year

- Table 3—Holdings: Information about the Party's holdings at the end of the year, providing a snapshot as of 31 December of the reported year

For countries hosting fluorocarbon project-based mitigation activities, much of this information will be collected by the government based on the information submitted by the project developer to the registry in use, and based on actions (including transfers, authorizations, uses, and cancellations) that may be carried out in the registry.

Regular Information

Countries must also submit regular information to the UNFCCC, as part of an annex to the BTRs.[70] Part of this information is similar to information submitted as part of the Agreed Electronic Format. Regular information will include information on how the party complies with the participation requirements of Article 6, including updates to any information submitted as part of an initial report or other BTRs. This will also include information on the authorization, such as information on the authorizations of the use of ITMOs for NDC use or Other International Mitigation Purposes, including any changes that may occur in the authorization and how corresponding adjustments were carried out and how ITMOs that have been used will not be further transferred, further canceled, or otherwise used.

They must also include information about each cooperative approach, including its ensuring of environmental integrity; measurement of mitigation co-benefits; and how each cooperative approach minimizes and where possible avoids negative environmental, economic, and social impacts. In line with this, each cooperative approach reported shall include a description of how it is consistent with national sustainable development objectives, safeguards and limits, and resources for adaptation if applicable, and deliver overall mitigation in global emissions if applicable.

Furthermore, the annex to the BTR will also include information about how human rights are maintained.

---

67   Intergovernmental Panel on Climate Change (IPCC). 2006. *IPCC Guidelines for National Greenhouse Gas Inventories.* https://www.ipcc-nggip.iges.or.jp/public/2006gl/.

68   UNFCCC, CMA. 2021. Decision 6/CMA.4, Annex VII, Matters Relating to Cooperative Approaches Referred to in Article 6, Paragraph 2, of the Paris Agreement. https://unfccc.int/documents/626570.

69   UNFCCC. 2024. Article 6.2 Reference Manual for the Accounting, Reporting and Review of Cooperative Approaches. https://unfccc.int/sites/default/files/resource/Article_6.2_Reference_Manual.pdf.

70   UNFCCC, CMA. 2021. Decision 2/CMA. Annex, Para. 21. Guidance on Cooperative Approaches Referred to in Article 6, Paragraph 2, of the Paris Agreement. https://unfccc.int/sites/default/files/resource/cma2021_10_add1_adv.pdf.

Importantly, countries must also include information on the metrics and measurement of the mitigation outcomes (Box 3). For projects transferring GHG units, such as HFCs, the country shall include a description of how each cooperative approach measures mitigation outcomes in accordance with the methodologies and metrics assessed by the Intergovernmental Panel on Climate Change (IPCC) and adopted by the CMA, the 2006 IPCC Guidelines.

The information described above is to be reported biennially as part of the regular information, through the structured summary, included in the BTR. The agreed upon common tabular format for the structured summary is available in Decision 5/CMA.3 (Annex II, Table 4).[71] The agreed upon common tabular format for the structured summary additionally contains the emission balance—where countries are to calculate their emission balance, representing the sources of emissions and removals covered by the NDC targets as well as the corresponding adjustments for ITMOs used or first transferred.

Accounting and Reporting of Ozone Depleting Substances

While ODS are a potent source of global warming, in addition to their ozone impacts, they are not covered under the Paris Agreement. Instead, they are regulated chiefly under the Montreal Protocol, which sets phaseout schedules of the gases for signatory countries.

Within a carbon market context, projects destroying or reclaiming ODS gases have been put in place in the past and continue to be developed and implemented. With Article 6, ODS projects may still be developed by countries if they ensure environmental integrity through appropriate baselines linked to national plans, strategies, and commitments under the Montreal Protocol, ensuring that the project is aligned with national plans, is additional, and has strong environmental integrity.

However, accounting of ODS projects based on the Paris Agreement is different from HFC gases, given that they are not the type of gas covered in the agreement. Broadly, Article 6 guidelines state that when a mitigation outcome is measured and transferred in non-GHG metrics, agreed upon by the participating countries, a description must be included in the BTR describing the method for conversion of non-GHG metrics (ODS) into tCO_2e.[72]

2.4.2 National Inventories and the Inclusion of Hydrofluorocarbons

National GHG inventories, as part of the Paris Agreement, are closely interlinked with Article 6. Countries rely on their measurement and calculation of national emissions, according to IPCC Guidelines, in order to track progress toward the achievement of NDCs. National inventories of GHG emissions are reported in the BTR.

Based on the modalities, procedures, and guidelines for the transparency framework under Article 13 of the Paris Agreement, the inclusion of HFC gases in national inventories is optional for developing countries.[73] Therefore, a number of countries do not currently report on HFCs or include them in their national inventory.

---

71 UNFCCC. 2021. CMA2021_L10a2: Report of the Conference of the Parties Serving as the Meeting of the Parties to the Paris Agreement on Its Third Session. https://unfccc.int/sites/default/files/resource/CMA2021_L10a2E.pdf#page=14.

72 UNFCCC, CMA. 2021. Decision 2/CMA.3, Annex, Para. 22(d), Guidance on Cooperative Approaches Referred to in Article 6, Paragraph 2, of the Paris Agreement. https://unfccc.int/sites/default/files/resource/cma2021_10_add1_adv.pdf.

73 UNFCCC, CMA. 2016. Decision 18/CMA.1, E. Reporting guidance, Para. 48. Modalities, Procedures and Guidelines for the Transparency Framework for Action and Support Referred to in Article 13 of the Paris Agreement. https://unfccc.int/resource/tet/0/00mpg.pdf.

Box 3

Metrics and Global Warming Potential

Decision 2/CMA.3 states that the participating Party is expected to communicate and report the metric to be used for generating and transacting internationally transferred mitigation outcomes (ITMOs). From these, two types of metrics are envisioned:

- Tonnes of carbon dioxide equivalent (tCO_2e) in accordance with the methodologies and metrics assessed by the Intergovernmental Panel on Climate Change (IPCC) and adopted by the Conference of the Parties serving as the Meeting of the Parties to the Paris Agreement (CMA).

- Other non-greenhouse-gas (GHG) metrics determined by the participating Parties that are consistent with the nationally determined contributions of the participating Parties.

Hydrofluorocarbon ITMOs may be reported and transacted as GHG metrics. When this is the case, the Party shall include information about how the measurement of the units has been carried out, including whether the ITMOs are being transacted in tCO_2e and how these have been measured and calculated in accordance with the methodologies and metrics assessed by the IPCC and adopted by the CMA (2006 IPCC Guidelines). This may include information on emissions factors such as those included in the 2006 IPCC Guidelines for Industrial Processes and Product Use, and conversion factors for units.

On the other hand, while having a global warming potential (GWP), ITMOs arising from ozone depleting substance (ODS) mitigation projects are not included as GHG metrics, and therefore are considered non-GHG metrics. For such projects, countries must report on how the cooperative approach ensures that the method for converting the non-GHG metric, such as an ODS, into tCO_2e, is appropriate for the context and mitigation scenario.[a]

One further important difference between the Paris Agreement and the Kigali Amendment to the Montreal Protocol are the GWP values that must be utilized for reporting under both treaties. The Paris Agreement currently employs GWP values arising from the IPCC Fifth Assessment Report (AR5, 2014), decided by the CMA.[b] On the other hand, the Kigali Amendment utilizes AR4 GWP values.[c] This may cause a difference in the values that are observed by a National Ozone Unit, and a unit tasked with reporting under the Paris Agreement. Broadly, Parties should be aware of the different values and how they may be addressed differently in the context of reporting under multiple treaties.

[a] UNFCCC, CMA. 2021. Decision 2/CMA.3 Annex, Paragraph 22(c), 22(d), Guidance on Cooperative Approaches Referred to in Article 6, paragraph 2, of the Paris Agreement. https://unfccc.int/sites/default/files/resource/cma2021_10_add1_adv.pdf.

[b] IPCC. 2014. *Climate Change 2013: The Physical Science Basis—Contribution of Working Group I to the Fifth Assessment Report of the Intergovernmental Panel on Climate Change.* https://www.ipcc.ch/site/assets/uploads/2018/02/WG1AR5_Chapter08_FINAL.pdf.

[c] IPCC. 2007. Changes in Atmospheric Constituents and in Radiative Forcing. In *Climate Change 2007: The Physical Science Basis—Contribution of Working Group I to the Fourth Assessment Report of the Intergovernmental Panel on Climate Change.* https://www.ipcc.ch/site/assets/uploads/2018/02/ar4-wg1-chapter2-1.pdf.

However, the modalities, procedures, and guidelines also include a provision in paragraph 48 to elaborate on the national inventory. The provision states that "any of the additional four gases (hydrofluorocarbons, perfluorocarbons, sulfur hexafluoride, and nitrogen trifluoride) that are included in the Party's NDC under Article 4 of the Paris Agreement, are covered by an activity under Article 6 of the Paris Agreement, or have been previously reported" shall be reported, regardless of flexibility provisions.

Therefore, when developing Article 6 activities in HFC, it is required that countries do include the "source and sink" of emissions into their national inventory, in this case, HFC gases.

2.4.3 Activities Outside the Scope of Nationally Determined Contributions

Countries may determine, independently and based on national circumstances, the sectors and emission sources and categories that they choose to cover as part of their NDC. However, as mentioned previously, for any Article 6 transaction, the gas type (including HFCs), must be included in their national inventory, based on the modalities, procedures, and guidelines under Article 13 of the Paris Agreement.[74]

This is not the case for the gases or sectors countries may choose to include in their NDC. No specific rule under the Paris Agreement that specifically requires countries to include a sector or emission category (such as the sector Industrial Processes and Product Use, in the case of HFCs) wherein Article 6 activities are taking place in the gases or sectors within their NDC's scope and coverage.

However, they are strongly recommended to do so, due to the potential for overselling. Otherwise, they face a situation where they must carry out corresponding adjustments for mitigation that is not being reflected as part of the target, as well as emissions accounting which counts toward the achievement of the NDC target (Figure 16).

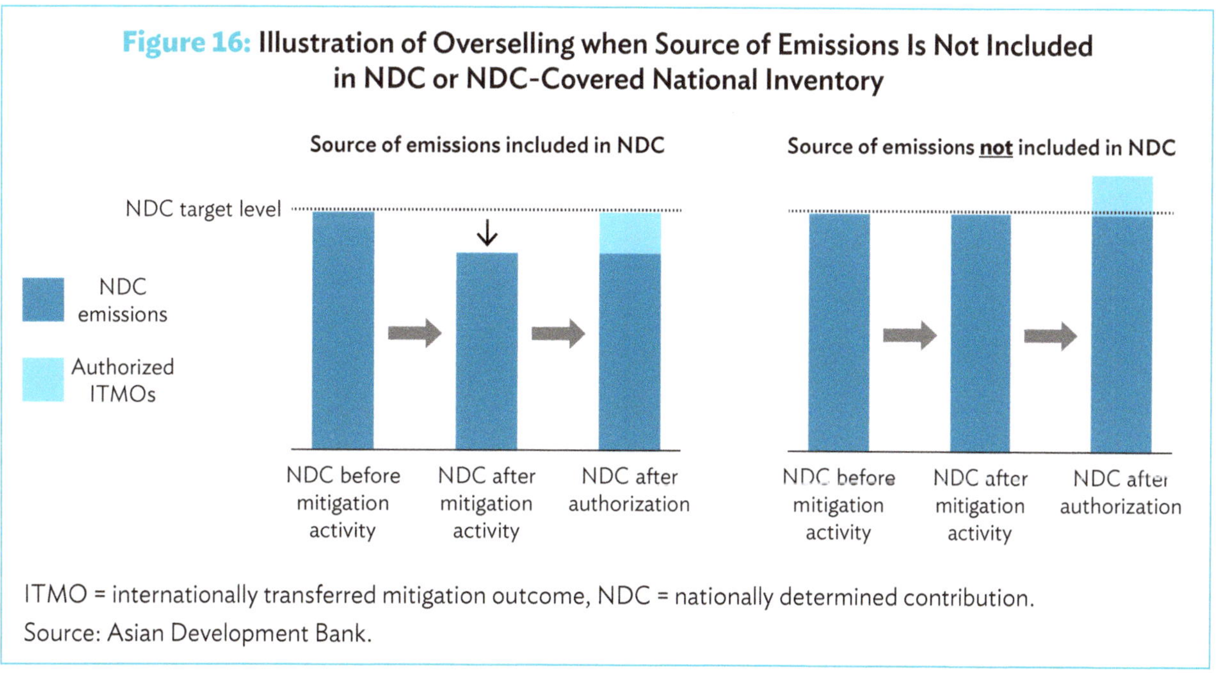

Figure 16: Illustration of Overselling when Source of Emissions Is Not Included in NDC or NDC-Covered National Inventory

ITMO = internationally transferred mitigation outcome, NDC = nationally determined contribution.
Source: Asian Development Bank.

[74] UNFCCC, CMA. 2016. Decision 18/CMA.1. Modalities, Procedures and Guidelines for the Transparency Framework for Action and Support Referred to in Article 13 of the Paris Agreement. https://unfccc.int/resource/tet/0/00mpg.pdf.

Countries that do not include the emission sector, source, or gas within their NDC will face the situation illustrated on the right side of Figure 16. The emission reductions that they achieve through a mitigation activity will not be reflected in the emissions accounting; therefore, any authorized ITMOs and corresponding adjustments will result in an upward adjustment toward their target without a matching reduction—essentially a "plus" without a "minus." This is often considered overselling, that is, the outward transfer of emission reductions, which are needed to achieve an NDC target.

Within a fluorocarbon carbon market project context, countries are therefore encouraged to include the sector (Industrial Processes and Product Use) and gas (HFC) within their NDC, if not done so already, ensuring that any emission reductions are reflected in the emissions balance and that any overselling risk is minimized.

CHECKLIST FOR LEVERAGING ARTICLE 6 INTERNATIONAL MARKETS FOR LIFE CYCLE REFRIGERANT MANAGEMENT

The following checklist provides an overview of the mandatory and optional but highly recommended steps that countries must take to leverage Article 6 of the Paris Agreement to support life cycle refrigerant management.

	Readiness Checkpoint	Is It Mandatory?	National Status
1	A comprehensive Article 6 framework is in place.	No—countries may flexibly establish procedures according to national circumstances but they must establish the required arrangements in their reporting obligations.	
2	Arrangements for authorizing under Article 6 are in place, including for cooperative approaches, authorizing participants and ITMOs for a use (NDC or OIMP).	Yes	
3	Arrangements for carrying out corresponding adjustments are in place.	Yes	
4	Initial report submitted to UNFCCC.	Yes	
5	Procedures for tracking, including for the entire project cycle and transfer, are in place.	Yes	
6	A registry tracking key project information as defined in Article 6 guidance has been adopted (gained access to) or established.	Yes	
7	Refrigerant gases (HFCs) are included in the scope and coverage of the NDC.	No—however, this is highly recommended in order to minimize overselling risks and leverage Article 6 effectively and transparently.	

continued on next page

continued

	Readiness Checkpoint	Is It Mandatory?	National Status
8	Refrigerant gases (HFCs) are included in the national inventory reported to the UNFCCC.	Yes—see paragraph 48 of MPGs.	
9	Comprehensive evaluation of national context.	No—however, coordinating with the National Ozone Unit, which may already have extensive knowledge, is highly recommended.	
10	Eligible carbon standards for project development have been determined.	Yes	
11	Bilateral agreements with trading countries have been signed.	Depending on the modality, use of ITMOs, and other arrangements, having in place finalized bilateral agreements may be mandatory.	
12	A GHG inventory of fluorocarbon gases according to IPCC guidelines has been developed.	Yes	
13	A comprehensive strategy for fluorocarbon management through carbon markets is in place.	No—however, this is recommended in order to establish arrangements and targets. Some countries may also choose to proceed with pilot or targeted projects to begin to understand capacity and national context.	

GHG = greenhouse gas; HFC = hydrofluorocarbon; IPCC = Intergovernmental Panel on Climate Change; ITMO = internationally transferred mitigation outcome; MPG = modalities, procedures, and guidelines; NDC = nationally determined contribution; UNFCCC = United Nations Framework Convention on Climate Change.

Source: Asian Development Bank.

www.ingramcontent.com/pod-product-compliance
Lightning Source LLC
LaVergne TN
LVHW071452180726
843512LV00018B/1358